HIERARCHIES'
ORIGIN

The Evolutionary Roots of Hierarchies

By
Covey Lee

Copyright 2024 by Covey Lee Every right reserved. No portion of this book may be reproduced in any form or by any electronic or mechanical means, including information storage and retrieval systems, without prior written permission from the publisher, with the exception of reviewers who may quote brief passages in their reviews.

Copyright © 2024

Table of Contents

Introduction:
Hierarchies' Pervasiveness

Part I:
The Environment

Chapter 1:
Understanding the Structured Complexity of Nature through Ecosystem Hierarchies

    1.1 Understanding Ecological Hierarchies in the Web of Life

    1.2 Trophic Levels: An Overview of Ecosystem Energy Flow and Ecological Hierarchy

    1.3 Keystone Species: Guardians of Biodiversity and Ecosystem Stability

Chapter 2:
Explaining the Origin and Diversities of Life via Evolution

    2.1 A Journey Into Deep Time: The Origins of Life and Order

    2.2 Elements of Evolution That Affect Hierarchy: The Creation of Structured Complexity

    2.3 The Role of Competition and Teamwork in Evolution and Society

Chapter 3:

The Structure, Function, and Significance of Animal Kingdom Social Hierarchies

    3.1 The Evolution, Structure, and Use of Dominance Hierarchies in Animal Societies

    3.2 Social Organization and Reproductive Strategies: The Associated Evolution of Behavior

    3.3 Case Studies: Examining How Primates, Wolves, and Other Animals Organize Socially and Use Reproductive Strategies

Part II:

The Dimension of Humanity

Chapter 4:

Tracking the Origins of Social Organization in Prehistoric Human Communities and Hierarchy

    4.1 Hunter-gatherer groups and egalitarianism: Analyzing Ancient Social Structures

    4.2 The Emergence of Social Complexity: Understanding the Path to Complex Human Societies

    4.3 How the Agricultural Revolution and Hierarchical Changes Transformed Human

Societies

Chapter 5:

The Hierarchies of Human History: An in-depth Examination of the Evolution of Social Structures

    5.1 The Rise of Empires and States: Impacting Human History

    5.2 Medieval Societies: An Intricate Web of Religious Hierarchies, Feudalism, and Monarchies

    5.3 A Revolutionary Era: The Industrial Revolution and Contemporary Societal Structures

Chapter 6:

Analyzing the Complex Networks of Human Behavior: The Psychological Basis

    6.1 The Human Brain and Social Hierarchies: Untangling the Neuroscience of Social Structure

    6.2 Social Hierarchy: The Dynamics of Dominance, Power, and Status

    6.3 The Pull Toward Social Equity: Collaboration and Aversion to Inequality

Part III:

Modern Views

Chapter 7:

Managing Authority and Influence in the Virtual World: Digital Age Hierarchies

    7.1 Digital Social Environment Navigation: Social Media, Virtual Communities, and Power

    7.2 The Influence of Technology on Hierarchies: From the Industrial Revolution to the Digital Age

    7.3 Virtual Organizations and Distributed Leadership: Managing the Future of Work

Chapter 8:

Issues that Cut Across Justice, Sustainability, and Inequality in the Modern World

    8.1 Analyzing the Relationship between Economic Inequality and Social Hierarchies

    8.2 Environmental Hierarchies and Sustainable Development: A Confluence of Preservation and Progress

    8.3 Challenging Hierarchies to Make the World Fairer: Social Justice Methods

Chapter 9:

Defining New Hierarchies and Power Structures for Social Change

    9.1 Going Beyond Traditional Power Structures: Reevaluating Governance and Leadership

9.2 Anarchism, Grassroots Movements, and Participatory Decision-Making: An Idea Convergence

9.3 Models of Community Governance and Indigenous Leadership: A Fusion of Tradition and Expertise

Chapter 10:

Organizational Futures: Adapting to a Changing Environment

10.1 Innovation and Resilience in the Twenty-First Century: Adjusting to a Changing World

10.2 The Vital Conflict: Balancing Hierarchy and Coordination

10.3 Towards Sustainable and Inclusive Hierarchies: A Framework for 21st-Century Governance

Conclusion:

Conclusions and Prospects: Choosing the Next Step

- Overview of Key Concepts: Understanding the Complexities of Hierarchies

- An Argument in Favor of Sustainable Hierarchies: Building an Ethical Community

Bibliography

List of references and further reading:

Expanding Your Hierarchy Knowledge
Index
   - Subject and Term Index: Working Around Knowledge Hierarchies

Introduction:
Hierarchies' Pervasiveness

Hierarchies are a fundamental aspect of human existence. Human interaction, decision-making, and perception of the world are all influenced by hierarchies. Social structures, the natural world, ecological order, and human societies are all home to them. In this introduction, we embarked on an exploration of the intricate realm of hierarchies, encompassing their historical foundations, purposes, and evolving significance in the contemporary era.

Putting Structures in Place

Essentially, a hierarchy is an arrangement of individuals, objects, or components based on their relative power, importance, or influence within an organizational structure. These hierarchical structures are ubiquitous and can take many different forms, like a nation's political structure, an organization's corporate ladder, or an ecosystem's food chain. Hierarchies are a vital tool for understanding and organizing the complex relationships that form our world.

Hierarchies in the Natural World:

Hierarchies are fundamental to natural systems. An ecological hierarchy can be seen, for instance, in the trophic levels of an ecosystem. The base is made up of plants and other producers, followed by herbivores and advanced carnivores. This hierarchical structure represents how nutrients and energy are distributed within the ecosystem. The maintenance of these hierarchies depends on keystone species due to their disproportionate ecological influence. The balance and stability of ecosystems depend on these organic hierarchies, which highlight the interdependence of species and the intricate relationships that exist within them.

Social Organizations:

In human societies, social hierarchies regulate the distribution of power, privileges, and resources. These hierarchies can be found in a wide range of contexts, including political structures, family and class divisions, and organizational charts. Social hierarchies can function in ways that are both empowering and restricting. While hierarchies can provide structure, accountability, and direction, they can also obstruct social mobility, prevent innovation, and perpetuate inequality.

The History's Development:

It's a complex tale of how hierarchies have changed and evolved over the course of human history. Flat organizational structures, shared decision-making, and an emphasis on egalitarian ideals were often features of early human societies. As societies evolved, hierarchies emerged, with leaders assuming positions of authority and control. Monarchies, empires, and feudal systems each represented distinct phases in the evolution of human hierarchies, with their own set of challenges and peculiarities.

The Contemporary Scene:

Even though hierarchies still have an impact on our world today, evolving paradigms are causing them to change. The Information Age gave rise to digital hierarchies, where information is a valuable resource and power and influence are typically decentralized. Global communication, remote work, and virtual organizations have all altered how we interact within hierarchical structures. There are advantages and disadvantages to these changes.

The Direction to Go:

As we examine the various dimensions of hierarchies, we encounter challenges and matters that require further examination. How can hierarchies adapt to the demands of a changing world and foster moral leadership, inclusivity, and sustainability at the same time? What lessons can history and science teach us about navigating the complexities of hierarchies in the digital age? These are the inquiries that direct our search for information.

In the pages that follow, we will look at the intricacies of ecological hierarchies, explore the dynamics of social structures, revisit the historical foundations of human hierarchies, and speculate about the future of organization and governance. Our research serves as a call to action to engage with the dynamic and ever-present world of hierarchies, to critically evaluate their role in our day-to-day existence, and to explore the possibilities for a future that is more equitable, sustainable, and inclusive.

Hierarchies affect every aspect of our life, from the ecosystems that sustain life to the societies that organize human activities. In our quest for knowledge and understanding, we will disentangle the complex web of hierarchies, trying to grasp their profound implications and envision a society in which hierarchies promote structure and order as well as justice, cooperation, and everyone's well-being.

Part I:

The Environment

Chapter 1:

Understanding the Structured Complexity of Nature through Ecosystem Hierarchies

Ecosystems, the intricate tapestries of life on Earth, are governed by a complex order that entwines different species and processes into a complex web of interactions. The foundation of this order is the concept of hierarchies. Because they affect the distribution of power, resources, and energy within these natural communities, hierarchies are essential to the structure and functioning of ecosystems. In this study, we delve into the fascinating field of ecosystem hierarchies, shedding light on their significance and the underlying ecological principles.

The Pyramid of Life's Trophic Levels and Energy Flow

One of the most fundamental hierarchies in ecosystems is the trophic pyramid, which shows how energy moves from one trophic level to another. The main process by which plants and other primary producers, including algae, get energy from the sun is called photosynthesis. They have reached the first trophic level, placing them at the base of the pyramid. Herbivores, or the consumers who eat these primary producers, are found in the second trophic level. Herbivore-eating predators comprise the third trophic level, and so on. Each trophic level corresponds to an increase in consumption and a decrease in available energy. This hierarchical structure controls how nutrients and energy are distributed throughout the ecosystem.

Keystone Species and Ecosystem Stability

Another way that hierarchies within ecosystems are expressed is through the concept of keystone species. Keystone species have a disproportionate amount of influence over their ecosystem despite their typically small numbers. By serving as pillars, they exercise control over the dynamics and composition of the community. The extinction of a keystone species could upset the entire hierarchy and have far-reaching consequences. In kelp forest ecosystems, the sea otter is a well-known example. Sea otters consume sea urchins, which are herbivores that graze on kelp. By managing the sea urchin population, sea otters indirectly contribute to the growth of kelp forests. This makes it possible for many other species to find food and habitat. Examples of how the hierarchical structure of ecosystems can support resilience and stability are keystone species.

Dynamics of Predator-Prey Relationships with Populations

The relationships between predators and prey, which have an impact on population dynamics, are another way that ecosystem hierarchies are shown. Predators use top-down control to influence the distribution and abundance of prey species. This top-down control affects multiple trophic levels across the hierarchy. For example, the reintroduction of wolves in Yellowstone National Park has a cascade of ecological effects. Wolf predation reduced the number of elk, which allowed vegetation to regrow. Because it offered food and cover, this was also advantageous to small mammals and birds. Predators, who are at the top of the trophic hierarchy, have a significant impact on the ecosystem as a whole.

Ecological Upheaval and Succession

Another process in which hierarchies are evident is the succession of ecosystems. Because of man-made or natural disturbances, ecosystems go through multiple developmental stages, each of which is linked to a unique hierarchical structure. For example, a barren landscape will undergo primary succession following a volcanic eruption, with pioneer species such as lichens and mosses settling there first. When soil and environmental conditions improve, more complex communities of plants and animals will eventually arise. The hierarchical development of ecosystems is demonstrated by this dynamic process.

Ecosystem hierarchies are not just theoretical concepts; they are the underlying framework for the organization of nature. These hierarchies determine how energy and resources move through ecosystems, influencing their resilience, structure, and overall functionality. These hierarchies may be connected to succession stages, keystone species, trophic levels, or predator-prey interactions. An understanding of these ecological hierarchies is crucial for the preservation and sustainable management of our natural environment because they provide guidance for how ecosystems operate and adapt to change. It demonstrates the complexity and interconnectedness of life on Earth and highlights how important it is that we take care of these intricate hierarchies in order to preserve ecological sustainability and harmony.

1.1 Understanding Ecological Hierarchies in the Web of Life

Ecosystems are the intricate networks of interconnected life on Earth. They are the result of millions of years of evolution, during which countless species interacted in a complex web of relationships. This web revolves around the concept of ecological hierarchies. To truly comprehend how ecosystems work, we must investigate the complexities of this intricate web of life and the hierarchical structures that underpin it.

Ecosystems as Complex Systems

Ecosystems, which are well-known examples of complex systems, are made up of many different species of plants, animals, microorganisms, and their physical surroundings interacting in a complex web of relationships. The essential element of this complexity is the hierarchical structure that regulates the flow of matter and energy within ecosystems. It is crucial to comprehend this hierarchy in order to understand how ecosystems function and adapt to change.

Trophic Stresses and Energy Transfer

One of the most fundamental ecological hierarchies is the trophic pyramid. Trophic levels are used to classify species based on their position in the food chain. At the base are primary producers, such as algae, plants, and photosynthetic bacteria. They convert sunlight into chemical energy through photosynthesis. Above them are herbivores, who consume the primary producers. Higher up are the herbivore-hunting carnivores, and so on. Every trophic level represents a step up the energy pyramid as energy descends from it.

Understanding trophic levels is crucial to comprehending how energy moves through ecosystems. A hierarchical structure governs the distribution of nutrients and energy, with energy decreasing as one climbs the pyramid. The relationships between species and the distribution of resources, which are impacted by this hierarchy, determine the structure and dynamics of ecosystems.

Principal Species of Biodiversity

Over the intricate web of life, some species have more influence than others. These species are known as keystone species because they are vital to maintaining the composition and efficiency of an ecosystem. They often comprise only a small fraction of the species in the ecosystem, but they have a disproportionate influence on the structure of the hierarchy.

In kelp forest ecosystems, the sea otter is a well-known example. Sea otters consume sea urchins, which are herbivores that graze on kelp. Managing sea urchin populations, sea otters indirectly promote the growth of kelp forests, which offer food and habitat to many other species. An excellent illustration of how ecosystems' hierarchical structure can support stability and biodiversity are keystone species.

Ecosystem Services and Human Well-Being

Significant effects of ecological hierarchies are seen on human well-being. Ecosystems provide a multitude of services that are vital to our existence and well-being. These services include sustaining nutrient cycles, managing the climate and preventing disease, supplying food and water, and offering cultural *(recreation and spiritual)* services.

The hierarchical structure of ecosystems ensures these services. For instance, the presence of keystone species aids in the regulation of ecosystem processes, and the trophic level structure ensures the availability of food resources. Understanding the importance of these services and the ways in which hierarchies affect their delivery is essential for sustainable human-environment interactions.

An understanding of ecological hierarchies is necessary to comprehend the intricate web of life in ecosystems. Trophic levels, biodiversity, and keystone species are all significant determinants of the resilience and functioning of these complex systems. Ecosystems' hierarchical structure affects the flow of energy, the distribution of resources, and the diversity of species found there. Ecological research and our ability to preserve and sustainably manage ecosystems are both dependent on this information, which will ensure human and natural world survival. As we navigate an era of environmental challenges, it is more important than ever to understand the web of life and its ecological hierarchies.

1.2 Trophic Levels: An Overview of Ecosystem Energy Flow and Ecological Hierarchy

Understanding trophic levels and the flow of energy within ecosystems is a fundamental aspect of ecology. It makes clear the intricate hierarchy that regulates the movement of energy and nutrients among the various organisms that make up an ecosystem. Essentially, the concept of trophic levels provides insight into the flow of life-sustaining energy from one organism to another and highlights the interdependence of species in the intricate web of life.

Determining the Trophic Phases

Trophic levels are a crucial element of the ecological hierarchy. They categorize the living things in an ecosystem according to their position in the food chain and their role in the exchange of energy. Trophic levels typically have a pyramidal structure at the base, containing primary producers, and at the top, herbivores, primary carnivores, and so on.

Primary Producers (Trophic Level 1): At the base of the trophic pyramid are the primary producers. These are autotrophic organisms that use photosynthesis to get energy from the sun. They mostly consist of algae, plants, and some bacteria. When solar energy is transformed into organic compounds, they produce carbohydrates, which are the main source of energy for all other creatures in the ecosystem.

Trophic Level 2 Herbivores: These organisms are positioned above primary producers. These are organisms that receive their food from primary producers. Examples include herbivorous mammals, grazers, and insects. Herbivores play a major role in the transfer of energy from primary producers to higher trophic levels.

Level 3 Tritrophic: Principal Carnivores: These organisms are the next trophic level; they feed on herbivores. These animals are known as primary carnivores or secondary consumers because they are carnivores that get their energy from herbivores. A few examples of predators are fish species, birds, and snakes.

Secondary Carnivores and Upward: At the base of the trophic pyramid are higher-order consumers, including tertiary, quaternary, and even apex predators. Every trophic level after it is represented by a smaller amount of energy as one climbs the energy pyramid.

Energy Transfer via Trophic Levels

Energy flow via the trophic levels is a fundamental ecological principle. As energy moves through an ecosystem, it becomes progressively less available at each trophic level. This decrease in available energy is caused by the second law of thermodynamics, which states that energy is lost as heat during each energy transfer.

The transfer of energy between trophic levels is one feature that sets ecological hierarchies apart. Primary producers absorb solar radiation, convert it into chemical energy, and store it as organic compounds. When herbivores consume primary producers, they release this stored energy. However, not all of the energy is transferred to the herbivores because some is lost as metabolic heat during digestion and respiration.

This process is at the top of the trophic pyramid. Once more, when primary carnivores eat herbivores, a significant portion of the energy stored in the former is lost as heat. Every subsequent trophic level experiences the same pattern of energy transfer and loss.

Resulting Effects on Ecological Dynamics

The hierarchical structure of trophic levels and energy flow has significant effects on ecosystem dynamics. It regulates the distribution of resources, population densities, and interspecies relationships. A rise or fall in population at one trophic level, for instance, could cascade into other trophic levels. These interactions are essential to the maintenance of ecosystem stability and the regulation of species diversity.

Understanding trophic levels and energy flow is crucial for ecologists and conservationists. By shedding light on the make-up and functioning of ecosystems, it helps us make decisions regarding the upkeep and conservation of these complex natural systems. By comprehending how trophic levels affect the dynamics of life within ecosystems, we can gain a better understanding of the interdependence of all living things and the importance of maintaining balanced trophic relationships for the health and sustainability of the various ecosystems that make up our planet.

1.3 Keystone Species: Guardians of Biodiversity and Ecosystem Stability

Ecosystems are intricate networks of life in which the existence of specific species has a disproportionate impact on the system's stability and diversity. These species are called *"keystone species"* because, like the keystone in an arch, their removal has the potential to bring the entire structure toppling. Understanding keystone species and their role in maintaining ecosystem stability is fundamental to the study of ecology.

Definition of Keystone Species

In the 1960s, ecologist Robert T. Paine made the term *"keystone species"* widely known. Keystone species are those that have an unexpectedly large impact on their ecosystem relative to their biomass or abundance. Their significance stems from their unique ecological roles, which have the capacity to drastically change the composition and structure of the entire ecosystem.

Ecologist builders and waterfalls

As one means of influencing others, keystone species are crucial to ecosystem engineering. They can physically change their environment, creating habitats for a diverse range of other species. For instance, beavers are renowned ecosystem engineers. Dam construction alters the water's flow, resulting in the creation of wetlands. These wetlands serve as vital habitats for a variety of aquatic plants, amphibians, insects, and waterfowl. Because they alter their environment, beavers have an indirect effect on the variety and abundance of other species in the ecosystem.

Keystone species are also capable of initiating Trophic cascades. Trophic cascades happen when changes in the abundance of a top predator or keystone species affect the whole food chain. For example, when gray wolves were reintroduced into Yellowstone National Park, a trophic cascade was created. Wolf predation caused elk numbers to decline. Because of this, elk had less opportunity to feed on willow and aspen trees, which helped them heal. The resurgence of these tree species brought with it changes to the composition of riparian habitats and to bird populations. The existence of wolves, as keystone species, triggered a chain of events that ultimately altered the ecosystem as a whole.

Preserving Biodiversity

In many cases, keystone species are essential to maintaining biodiversity. By altering the abundance and distribution of other species, they can stop the dominance of a single species or a small group of species. The ecosystem grows more varied and resilient as a result. In the absence of keystone species, certain species may become excessively abundant to the extent of displacing other species and reducing biodiversity in general.

Kelp forest ecosystems offer a well-known example of this concept. Keystone species in these ecosystems, sea otters, eat sea urchins, which are herbivores that graze on kelp. By controlling the sea urchin population, sea otters indirectly promote the growth of kelp forests, which provide food and habitat for a variety of other species, including fish, invertebrates, and even marine mammals.

Challenges and Maintenance

The importance of keystone species to the health of ecosystems emphasizes the need for their preservation. Unfortunately, many keystone species are in danger as a result of human activities like habitat destruction, overexploitation, and climate change. The extinction of a keystone species can have cascading effects on ecosystems as a whole.

It is necessary to protect and restore keystone species and their habitats for the sake of biodiversity and ecological balance. Protection programs for top predators like wolves, apex predators like sharks, and ecosystem engineers like beavers, for example, can significantly improve the ecosystems in which they reside.

The unsung heroes of biodiversity and ecosystem stability are keystone species. Their influence in an ecosystem extends well beyond their numerical representation; they influence the intricate web of life through trophic cascades, habitat modification, and species diversity regulation. Understanding and protecting these keystone species is essential if we are to maintain and restore the resilience and health of the natural world as well as ensure that ecosystems support the various forms of life that depend on them.

Chapter 2:

Explaining the Origin and Diversities of Life via Evolution

The concept of evolutionary roots lies at the heart of modern biology. This clever framework helps us understand the origin, diversity, and interconnectedness of life on Earth. Genetics and other scientific disciplines have contributed to the theory of evolution since it was first put forth by Charles Darwin. Collectively, they provide a comprehensive understanding of species evolution and adaptation.

Darwin's Theory of Evolution Based on Natural Selection

Charles Darwin's theory of evolution, first presented in his landmark work ***"On the Origin of Species"*** in 1859, popularized the idea that species change over time through a process he termed *"natural selection."* Natural selection is essentially a process that aids in a species' ability to change and adapt to its environment.

It operates based on several core principles:

Variation: Within a population, there is genetic variation. No two people are exactly the same.

One generation inherits and passes on certain traits to the next. Children inherit traits from both parents in combination.

Competition: Living things compete with one another for limited resources, such as partners, food, and shelter. Not everyone is here to procreate.

Adaptation: Individuals with traits that better suit their surroundings have a higher chance of surviving, procreating, and passing those traits on to their offspring.

Over many generations, natural selection can significantly alter a population, which may ultimately result in the emergence of new species. This process is the cornerstone of evolutionary biology.

Genetics and Molecular Evolution

The framework for understanding the *"how"* of evolution was supplied by Darwin's theory of evolution; however, the *"why"* and the mechanisms underlying it were clarified by the discovery of genetics. Gregor Mendel's work on the laws of inheritance in the late 19th century and the subsequent development of molecular genetics in the 20th century provided significant insights into the genetic foundations of evolution.

Genetics has shown that changes in DNA sequences, which are passed down from generation to generation, are the cause of species variations. Genetic recombination, mutations, and drift are the causes of these variations. We can now track evolutionary relationships and create phylogenetic trees, which show the genetic relationships between different species, thanks to molecular biology. By comparing DNA sequences, we can gain insight into the evolutionary history of life on Earth, our common ancestry, and the mechanisms underlying speciation.

The Tree of Life and Biodiversity

Our understanding of the diversity of life is significantly influenced by genetic discoveries and the theory of evolution. The *"tree of life"* concept illustrates how all species share a common ancestor. By exposing the evolutionary connections between all living things, it demonstrates the interconnectedness of all life forms.

This relationship goes beyond simple biological classification. It also highlights how important biodiversity is to ecosystem stability and health. The great diversity of life on Earth provides a vast reservoir of ecological and genetic resources. Apart from guaranteeing that ecosystems can adapt to evolving conditions, biodiversity provides vital services for human survival such as pollination of crops, uncontaminated water and air, and regulation of climate.

Issues and Opportunities for the Future

The evolutionary underpinnings of biology are well established, but ongoing studies continue to uncover novel aspects of evolution, including epigenetics and horizontal gene transfer. In addition, contemporary environmental issues such as habitat loss, climate change, and the sixth mass extinction event emphasize how important it is to understand evolution in order to address environmental issues on a global scale.

The foundation of evolution is one of the biological sciences' main topics. They provide a comprehensive framework for understanding species interdependence, the diversity and origins of life, and the importance of biodiversity for the well-being of our planet. Evolutionary biology is still a vibrant field of study that guides our efforts to protect Earth's remarkable diversity while illuminating the origins and mechanisms of life.

2.1 A Journey Into Deep Time: The Origins of Life and Order

From the primordial soup of Earth's early history to the complex and hierarchical ecosystems we see today, the story of life's origins spans billions of years. The emergence of hierarchical structures within the universe and the processes leading up to it are fascinating subjects that blend biology, chemistry, and geology. To comprehend how life evolved and gave rise to the intricate hierarchies we see today, we must delve into the depths of deep time.

From Molecular Replication to Archaic Soup

The origins of life on Earth are still a topic of debate and research for scientists. One of the most well-known theories, the primordial soup hypothesis, suggests that life first began in a nutrient-rich organic compound *"soup"*. Under the right conditions, these materials could have evolved into the building blocks of life, such as lipids, nucleotides, and amino acids.

This early life was probably made of hierarchically arranged self-replicating molecules. These molecules may have evolved and diversified over time to give rise to the complexity we see in modern organisms. The ability to replicate itself was a crucial step in the hierarchy of life since it prepared the groundwork for the inheritance and storage of genetic information.

The Growth of Living Cells

Under the hierarchy of life, the next significant development was the emergence of cellular structures. The transition from simple organic molecules to self-contained cells represents a turning point in the evolutionary history of life. It enabled organisms to develop greater specialization and organization.

All multicellular organisms, including plants, animals, and fungi, are considered to be composed of eukaryotic cells, which are believed to have emerged from the process of endosymbiosis, in which two cells engulfed one another and formed a symbiotic relationship. Eukaryotic cells are characterized by membrane-bound organelles like the nucleus and mitochondria that perform particular functions inside the cell.

Multicellular Organisms Emerging from Unicellular Species

Eukaryotic cells could appear and lead to the development of multicellular organisms. The transition to multicellularity marked another significant turning point in the life hierarchy. In multicellular organisms, the division of labor among specialized cells and tissues enables them to perform a greater variety of tasks and adapt to a greater variety of environments.

The evolution of multicellularity gave rise to the wide variety of complex life forms that we see around us, from the massive rainforest trees to the intricate social insect societies. In all cases, the organisms themselves display hierarchies, with various functions performed by the tissues, cells, and organs to ensure the survival and propagation of the whole.

Natural Hierarchies and the Life Cycle

Ecological hierarchies arise when we shift our focus from individual organisms to ecosystems. These hierarchies are defined by trophic levels, the points at which energy transfers from primary producers to herbivores, carnivores, and decomposers. Each level represents a particular stage in the ecosystem's flow of nutrients and energy.

Ecosystem stability and hierarchy depend on keystone species. Similar to the origins of life, the emergence of keystone species can be understood as a turning point in the hierarchy of ecosystems. They affect the distribution of resources, the diversity of species, and the overall stability of ecosystems.

Challenges and Ongoing Research

Researchers continue to look into the origins of life and the development of hierarchical structures within it. Finding out if extraterrestrial life exists and, if so, what kind of life it could have had, is the goal of the study of astrobiology.

As we learn more about the origins of life and the hierarchy on Earth and beyond, we gain a better understanding of the intricate web of life that surrounds us. This knowledge not only informs our perspectives on sustainability and conservation, but also aids in our understanding of evolution and ensures that life's hierarchical structures remain robust and resilient in the face of environmental challenges.

2.2 Elements of Evolution That Affect Hierarchy: The Creation of Structured Complexity

Evolution is intimately linked to the concept of hierarchy, which can be seen in everything from animal social structures to ecosystem configurations. The development of biological systems has been impacted by a multitude of factors, which has led to the emergence of hierarchical structures. Understanding the complexity and variety of expressions of life requires an understanding of the evolutionary forces underlying hierarchy.

Competition for Limited Resources

One of the primary evolutionary forces underlying hierarchy is competition for limited resources. Throughout the history of life on Earth, animals have had to contend with one another for access to necessities such as food, water, shelter, and partners. This competition has led to the emergence of hierarchies, in which some individuals or groups outbid others for limited resources. In the kingdoms of plants and animals, hierarchies of this type are common.

Within a forest canopy, for example, taller trees may outcompete shorter ones for sunlight. This results in a hierarchy of trees. Dominance hierarchies can form among animal communities, with the most dominant individuals having first dibs on resources like food or mate for reproduction.

Natural Selection and Reproduction's Success

In the natural world, two other significant factors that drive hierarchy are natural selection and the success of reproduction. Members of a species that display traits that augment their ability to reproduce and transfer genes are generally more dominant and influential in their respective populations.

For example, in many animal species, the most dominant individuals also usually have the highest success rate in mating and producing offspring. As a result, hierarchical social structures may eventually emerge, giving some individuals or groups the upper hand when it comes to procreating and genetically transferring traits to the following generation.

Predation and Predator-Prey Dynamics

Predation is one of the key factors affecting hierarchy in ecosystems. The distribution and abundance of prey species are significantly impacted by predators. Because they are typically found at the top of the food chain, predators have a top-down effect on prey populations.

Predator-prey dynamics can give rise to hierarchical structures in which certain species or individuals become dominant predators, while others at lower trophic levels become prey. The general hierarchical structure of ecosystems is supported by the relationships that influence the abundance and behavior of predators and prey.

Cooperative and giving behavior

Even though competition and predation can still be the primary motivators behind hierarchical structures, cooperation and altruism within individuals or groups also play a significant role in the evolution of hierarchy. Cooperative behavior can help people in ways that improve their chances of surviving and raising successful offspring.

Termites, ants, and bees are examples of social insect colonies where different individual castes work together to form complex societies with division of labor. These colonies are ruled by cooperative behavior that benefits the colony as a whole. They have a queen, worker castes, and a hierarchical structure.

Adjusting to Environmental Shifts

Hierarchies of biological systems are also influenced by changes in the environment and the adaptability of organisms. Over time, environmental conditions shift, forcing species to adapt or face extinction. The ability to adapt to different ecological niches and make use of new resources can lead to species diversity and hierarchical structures within ecosystems.

For example, species that can adapt to new environmental conditions and successfully occupy a niche in a newly formed habitat often become dominant in that ecosystem. This process facilitates the hierarchical distribution of species in diverse habitats.

Challenges and Opportunities for the Future

Understanding the evolutionary processes underlying hierarchy can aid in a deeper comprehension of the organization of biological systems. However, it's important to realize that hierarchy is a concept that is flexible and not applicable to all situations. Different ecological contexts give rise to different levels of hierarchical organization in different species and ecosystems.

As we explore the complexities of hierarchy in biology, our understanding of the interdependence of all life and the dynamic processes that have given rise to organized complexity on Earth has expanded. Understanding these evolutionary processes also helps us to appreciate the ecological roles that hierarchy plays and how crucial it is to maintaining the stability and equilibrium of natural ecosystems. It is proof of the remarkable adaptability and diversity of life as well as the processes of evolution that have shaped the biological world into what it is now.

2.3 The Role of Competition and Teamwork in Evolution and Society

In human society and biology, cooperation and competition are two fundamental forces. They affect the behavior of organisms, the evolution of species, and the structure of human communities. In social as well as natural contexts, these seemingly opposing forces often work together to create complex hierarchies and systems. Comprehending the functions of cooperation and competition is essential to comprehending the dynamics of life on Earth.

Collaboration in the Natural World

In the natural world, organisms cooperate with one another and occasionally even with different species. Mutualism, a sort of cooperation in which the interactions of two species are beneficial to both, is one of the most prominent examples. For instance, there is a mutualistic relationship between bees and flowering plants in which the plants give nectar to the bees and the bees pollinate the plants. In ecosystems, this type of mutually beneficial cooperation is common and facilitates successful reproduction for both sides.

Cooperation is also seen in social animals like wolves and dolphins, where group living and cooperative hunting increase survival rates. Social insects, like ants and bees, display a highly structured type of cooperation that includes coordinated efforts and the division of labor among colony members.

There is a definite evolutionary advantage to cooperation. When organisms work together, they can frequently fend off predators, get resources faster, and increase the overall fitness and survival rate of the group. Complex social structures and social behavior have developed through collaboration.

Competition in Nature

In nature, cooperation and competition often coexist and serve as motivating factors. When organisms battle it out for limited resources, like food, water, territory, and mates, both within and between species, competition takes place. The struggle for survival and the right to procreate leads to competitive behaviors and traits.

Competition can result in the survival of the fittest because individuals with advantageous traits have a higher chance of obtaining resources and passing on their genes. It is the main driving force behind natural selection and other evolutionary processes that have shaped Earth's diversity of life. Individuals and species need to adapt to shifting environments and fight for a place in the ecological hierarchy.

Predation is another aspect of competition in ecosystems. The dynamics of the competitive relationship between predators and their prey influence the distribution and size of populations within ecosystems. This rivalry contributes to the hierarchical and complex nature of ecological systems.

Harmonizing Human Society's Rivalry and Collaboration

In human society, cooperation and competition are essential elements of economics, culture, and social structure. Rather than being mutually exclusive, these two forces often coexist and interact in intricate ways.

Cooperation is essential for the growth of social bonds, group projects, and the creation of shared norms and values. Humans have developed complex organizations, technologies, and societies due to our ability to work together on a large scale. Cooperation can be seen in fields such as science, where scientists from different backgrounds work together to solve difficult problems, and in the global economy, where trade and cooperation between nations are the primary forces behind economic growth and development.

On the other hand, competition promotes economic progress, inventiveness, and individual success. It encourages individuals and groups to strive for excellence, which frequently results in the development of novel goods, improved technologies, and more efficient organizational structures. In economics, market competition can lead to more reasonable costs and better products, which is advantageous to consumers.

Finding a balance between cooperation and competition is a never-ending challenge in human society. Excessive competition can lead to social inequality and disparities, whereas excessive cooperation can stifle innovation and economic advancement. Achieving the right balance is crucial for society's progress and well-being.

Instead of being two opposing forces, cooperation and competition are two sides of the same coin that shape human society and drive evolution. Diverse species and intricate ecosystems have emerged as a result of the complex interactions between these forces. These interactions also play a fundamental role in human society, providing the framework for social structures, economic systems, and cultural dynamics. Understanding the roles that competition and cooperation play in our understanding of life and society is essential for navigating the complexities of our world and promoting positive and sustainable outcomes.

Chapter 3:

The Structure, Function, and Significance of Animal Kingdom Social Hierarchies

All animals have social hierarchies, including fish, mammals, birds, and insects. These hierarchical structures frequently dictate how members have access to resources, opportunities for mating, and general social dynamics, which has an impact on how members of a group interact and behave. Understanding animal social hierarchies provides important insights into the emergence and functioning of complex social structures.

Place and Dominance

The majority of animal social hierarchies are founded on dominance relationships, in which group members hold different ranks. The establishment of dominance often occurs through confrontational interactions, such as physical altercations or violent outbursts. Dominant individuals hold a higher rank in the hierarchy and are privileged to have access to resources and potential partners.

Some species have linear dominance hierarchies in which specific group members either surpass or are surpassed by individual members. In other cases, hierarchies become more complex, resulting in networks where individuals hold different levels of dominance in various social contexts.

Obtaining Resources

One of the primary functions of social hierarchies in the animal kingdom is to regulate access to essential resources. These resources could include chances for procreation, food, drink, and housing. Dominant individuals often have priority access to these resources in order to maximize their chances of survival and successful reproduction.

For example, in a pride of lions, the dominant males have better access to food and hunting spots. In a similar vein, dominant members of a group of primates may get first choice when it comes to mates, which increases their likelihood of successful reproduction.

Division of Labor

A division of labor is closely linked to hierarchical structures in some social species, most notably in social insects like termites, ants, and bees. Individuals in these colonies are born into discrete castes, each with specific responsibilities. The division of labor is necessary for the colony to survive and function.

In a hive of honeybees, for instance, worker bees perform varying duties based on their age and caste, such as foraging, nursing, and hive protection. The colony uses its resources as efficiently and productively as possible thanks to the division of labor.

Putting an End to Conflicts

Social hierarchies within animal groups also influence the ways in which disputes are settled. Dominance hierarchies can maintain peace within the group and lessen the chance of physical confrontations. It's common to use rituals and social cues to establish dominance without using physical force.

For example, wolves establish hierarchy and send messages through ritualized displays of dominance and submission that include body postures and vocalizations. These behaviors lessen the likelihood of injuries occurring within the pack by minimizing the need for direct physical confrontations.

Arrangement and Consistency

When animal groups have social hierarchies, they are more cohesive and stable. Their ability to maintain order and minimize chaos makes it possible for people to collaborate. Hierarchies are commonly used to ensure that group members can collaborate effectively and to reduce internal competition.

For example, dominant individuals in a flock of birds may lead the group during migration or foraging in order to keep the group cohesive and in order. This coordinated behavior increases the group's chances of survival and overall success.

Flexibility and Flow

Social hierarchies are dynamic and subject to change as a result of external factors, shifts in life stages, or adjustments in personal dominance. This flexibility allows groups to adapt to changing circumstances and the availability of resources.

In certain species, challenges to the established hierarchy may result in changes in rank. When new people challenge and replace established ones, the hierarchy might be rearranged. Social groups' resilience in changing environments rests on their capacity for adaptation.

Complexity and Adaptation

Animal social structures are complex, multifaceted systems. They carry out a variety of duties, including managing the allocation of work and resource access, lowering tensions, and planning group activities. These hierarchies are the result of millions of years of evolution, in which people have shaped cooperation and competition by trying to secure the success of their groups and maximize their own fitness. Studying social hierarchies in the animal kingdom offers a window into the complexity of social behavior and the adaptability of species in a range of ecological settings.

3.1 The Evolution, Structure, and Use of Dominance Hierarchies in Animal Societies

Animal societies often exhibit dominance hierarchies, wherein individuals within a group establish and maintain roles of either subordination or dominance. Controlling social interactions, resource access, and group dynamics all depend on these hierarchies. Dominance hierarchies vary amongst species and provide valuable insights into the evolution of social behavior because they are a reflection of their unique social structures and ecological environments.

Structure of Dominance Hierarchy

Depending on the species and social context, dominance hierarchies can have a linear or non-linear structure. Each individual in a linear hierarchy is positioned in a distinct, linear hierarchy, with those at the top ranking lower than those at the bottom. Hierarchies in animals like wolves are frequently linear, with pack members occupying specific positions based on their dominance.

Non-linear hierarchies, on the other hand, are more complex and resemble networks. People may be ranked differently in these hierarchies according to the specific resource or context they are competing for. Non-linear hierarchies, for instance, are commonly observed in social structures that are more flexible and pliable in some primates.

How to Achieve and Maintain Dominance

Establishing dominance in a hierarchical structure often involves competitive interactions such as physical altercations, violent outbursts, or ritualized actions. Dominance in a species can be determined by its distinctive traits, such as age, sex, or physical prowess. It is not solely determined by aggression; it is also influenced by one's ability to maintain social relationships, exhibit cooperation, and navigate difficult social situations.

Many species constantly assess and reinforce their dominance through a variety of behaviors, including posturing, vocalizations, and subtle signals. By reducing the need for physical altercations, these interactions preserve social order and reduce the risk of injury.

The Goal of Dominance Hierarchies

Dominance hierarchies serve several essential functions in animal societies.

Distribution of Resources: Access to resources, such as food, partners, and shelter, is governed by dominance structures. Dominant individuals are usually given priority access because it increases their chances of surviving and procreating.

Resolving Conflicts: Internal conflicts are less frequent and less severe when organizational structures are in place. Those with distinct dominance relationships are better able to maintain social order and prevent needless confrontations.

Division of labor: In some species, there is a caste system or hierarchy that assigns specific roles and responsibilities to members of the group. This increases group output and makes the best use of the resources at hand.

Reproduction: Dominance often determines access to mating opportunities. When dominant individuals monopolize reproductive partners in animals, the hierarchy has a major effect on the genetic makeup of the population.

Coordination: Maintaining order and arranging group activities are made easier by dominance hierarchies. They provide a framework for group behavior and ensure that the group runs smoothly.

The Significance of Evolution

Dominance hierarchies are the result of natural selection and are impacted by the unique ecological and social issues that different species encounter. Hierarchically organized groups are more capable of operating efficiently and cooperatively, which enhances the group's general fitness.

The evolution of dominance hierarchies is intimately linked to the development of complex social systems and social behavior. In species where these traits are critical for survival, dominance hierarchies help to facilitate coordinated and cooperative behaviors. Over time, species have evolved a variety of hierarchical-building and maintenance strategies, depending on their social structures and ecological niches.

An overview of social dynamics

The intricate social dynamics of animal societies are evident in the dominance hierarchies that have arisen across the animal kingdom. These hierarchies shed light on how individuals navigate their social settings in order to improve their chances of surviving and procreating. They also show the way in which competition and cooperation interact. An understanding of the composition, evolution, and role of dominance hierarchies is necessary to comprehend the diversity of social behavior and the adaptability of species in a range of ecological settings.

3.2 Social Organization and Reproductive Strategies: The Associated Evolution of Behavior

Reproductive strategies and social structures in animal behavior are closely related. The social structure of an animal often affects its reproductive habits, and vice versa. These dynamics are influenced by ecological environments and evolutionary pressures, which lead to the emergence of a vast array of strategies in the animal kingdom.

Social Structure: Moving From Independent to Collaborative Living

Animal social organization can take many different forms, ranging from complex group structures to lone individuals. Every species has evolved a social structure that best suits its ecological niche and reproduction requirements.

Solitary Species: Animals classified as solitary are those that lead primarily independent lives and socialize primarily for the purpose of procreation. This includes a wide range of reptiles, such as snakes and some lizard species, as well as some mammals, such as the lone cat species. Living alone can be advantageous in situations where resources are scarce and dispersed because it lessens competition for those resources.

Certain animals form monogamous pairs in which the male and female develop a close bond and collaborate to raise their young. This behavior is known as pair-bonding. This kind of social organization is commonly seen in some bird species, such as albatrosses and swans, and in some mammalian species, such as gibbons and beavers. When parents work in pairs, they typically alternate in caring for and supporting their children.

Living in Groups: Many species, particularly mammals like wolves, elephants, and primates, are gregarious. There are several advantages to living in a group, such as improved protection from predators, enhanced social learning, and enhanced foraging efficiency. Reproductive strategies may be impacted by the roles and hierarchies that members of these groups typically occupy.

Reproductive Methods: Comparison of Polygamy and Monogamy

Animal reproductive strategies differ greatly and are influenced by social hierarchy. Promiscuity, polyandry, polygyny, and monogamy are some of these strategies.

The formation of a dedicated, long-term pair bond between a single male and a single female is known as monogamy. The child is raised in part by both spouses. Monogamy is most common in species where offspring survival depends on parental care. For example, monogamous bird species share common responsibilities such as tending to the eggs and feeding the young.

A reproductive tactic known as polygyny involves a single male mating with multiple females. This strategy is commonly used by males to entice and mate with multiple females in species where they have territorial or resource control. Anima n ds include elk and some species of seal that live in harems.

Polyandry, in which a single female mates with multiple men, is the opposite of polygyny. This strategy is relatively rare, but it does exist in species where it can help females to have multiple males to help care for and raise their young. These species are also endowed with an abundance of resources. Examples include certain species of frogs and seahorses.

People that use promiscuity as a reproductive strategy mate with multiple partners, often without forming committed relationships. This strategy is common in species where access to mates is highly variable and sexual competition is intense. Males may engage in competitive mating rituals in many insect species where promiscuous mating is the norm.

Adaptable Trade-Offs

When choosing a specific social structure and reproductive strategy, there are frequent trade-offs. Living in a group, for instance, can improve social learning and protection, but it can also lead to more competition for resources and mating opportunities. Monogamy can limit a person's capacity for reproduction even though it might enhance parental care. Polygyny can give access to multiple partners, but it can also increase male-to-male competition.

Trade-offs like these are influenced by ecological factors like the distribution of mates, predation pressure, and resource availability. Evolution has favored strategies that maximize offspring survival and reproductive success within a given ecological context.

An intricate exchange

The diversity and adaptability of life on Earth are demonstrated by the interplay between reproductive strategies and social organization. These behaviors have developed as a result of the ecological niches into which these species have evolved, as well as the challenges they face in securing food, warding off predators, and successfully reproducing. Understanding the complex interplay between reproductive strategies and social organization is essential to appreciating the wide variety of behaviors exhibited by animals across the globe. It shows how ecological dynamics and evolutionary processes are intricately related, and how this has resulted in the extraordinary diversity of life.

3.3 Case Studies: Examining How Primates, Wolves, and Other Animals Organize Socially and Use Reproductive Strategies

To gain a deeper understanding of the complex relationship between social organization and reproductive strategies in the animal kingdom, we can examine case studies of a wide range of different species. These include primates, wolves, and other fascinating animals that show the intricate connection between social structure and reproduction.

First Case Study: Chimpanzees, or Pan troglodytes

Chimpanzees are social primates and some of our closest relatives. They live in societies with mixed-gender populations and complex social structures. Males compete in a community for supremacy, and the top-ranked males often have better access to mating opportunities. Chimpanzees, on the other hand, have a promiscuous mating system whereby males and females engage in sexual activity with multiple partners. Males are less likely to target unrelated infants because it is difficult to determine a female's paternity, which may help reduce the risk of infanticide. Additionally, coalitions, cooperative hunting, and enduring bonds are crucial elements of their social structure.

Case Study 2: African Elephants (Loxodonta africana)

In the matriarchal society that African elephants live in, the dominant female, referred to as the matriarch, sets group policies and directives. This social structure is based on a matrilineal hierarchy, in which related females form stable family units with their offspring. Elephant males usually live alone or temporarily form bachelor groups after leaving their birth groups. Male elephants compete with estrous females for their attention through polygynous mating, a reproductive tactic. The matriarch's extensive knowledge of resources makes her vital to the group's survival and future generations.

Case Study 3: Gray wolves, or Canis lupus

Gray wolves exhibit a complex social structure based on a pack structure. Packs are made up of alpha breeding pairs, which are made up of a dominant male and female, as well as their offspring and occasionally unrelated individuals. The dominant pair usually gets all of the pack's mating opportunities when it comes to reproduction. However, not every member of the pack bears children; many individuals assist in raising the offspring of the alpha pair. Using this cooperative breeding method, puppies gain more resources and live longer. Like wolves scattering, occasionally unrelated individuals may also establish new packs and generate their own chances for reproduction.

Case Study 4: Honey Bees, or Apis mellifera

Honeybees are a highly social insect species that rely on labor sharing to keep the colony alive. A colony of honeybees consists of a single fertile queen, male drones, and sterile female worker bees. The queen, the only reproductive female in the colony, can lay thousands of eggs each day. Worker bees are responsible for feeding, foraging, and maintaining the hive. Drones exist primarily to mate with queens. The efficient operation and proliferation of the colony are ensured by the division of labor among highly specialized individuals.

Case Study 5: African Lions, or Panthera leo

Prides are hierarchically structured groups of related females, their young, and a coalition of males that form the foundation of African lion societies. In a pride, dominant males maintain control over mating opportunities. Male coalitions often form in bachelor groups before confronting resident males for pride and mating rights. Female lions in a pride coordinate their estrous cycles to produce offspring as a group. The two main roles of the hierarchical social structure—which is inhabited by males and dominated by females—are cub raising and territory defense.

Diversity in Social Organization and Reproductive Methods

These case studies highlight the striking diversity of reproductive strategies and social structures found throughout the animal kingdom. Different strategies have been evolved by species to address the challenges posed by resource availability and ecological niches. The interplay between social structures, reproductive patterns, and ecological environments shows how complex and adaptable life is on Earth. By looking at these case studies, we can discover a lot about the intricate relationships that exist between behavior, evolution, and animal survival strategies.

Part II:
The Dimension of Humanity

Chapter 4:

Tracking the Origins of Social Organization in Prehistoric Human Communities and Hierarchy

As early human societies are studied, hierarchy—a crucial element of human social structure—emerged. The hierarchical structure of early human groups from the Paleolithic era was crucial to understanding the functioning of these communities. To understand the origins and functions of hierarchy in early human societies, we need to look at the archaeological and anthropological evidence from different periods of human history.

Paleolithic Era: Clades and Groups

During the Paleolithic era, some 2.6 million years ago to 10,000 years ago, early humans lived in small groups known as bands or tribes. These were usually nomadic groups that subsisted mostly on hunting and gathering. They showed some basic social differentiation in addition to simple hierarchical structures.

In bands, leadership roles were often determined by age, experience, or specialized knowledge. The group's decisions and actions were largely guided by senior and experienced hunters. This early hierarchy could be seen as a response to the need for leadership and collaboration in small, close-knit communities.

These hierarchies were most likely flexible and situation-specific, with leaders recognized for their expertise in particular domains.

The Neolithic Revolution: The Rise of Agricultural Societies

The Neolithic Revolution, which began around 10,000 years ago and saw a transition from hunting and gathering to agriculture and settled life, marked a significant turning point in human history. This change resulted in the emergence of more complex societies with higher population densities. In these farming communities, hierarchical structures were valued more highly.

A result of the agricultural surplus was a specialization in labor and the emergence of distinct social classes. Elites or powerful rulers arose because a small group of people possessed the resources and excess food. Based on resource control, this early hierarchy was defined by a social division between laborers and landowners. Establishing power and direction was essential for overseeing growing communities, distributing resources, and settling conflicts.

Ancient Civilizations: Complex Social Structures and Class Structures

Human societies evolved into increasingly complex hierarchies. Ancient Mesopotamia, Egypt, and the Indus Valley all had clearly defined social classes and extremely organized hierarchies. The highest positions in these hierarchies were held by rulers, priests, and nobles who had both political and religious authority. The lowest group consisted of slaves and servants, with merchants, laborers, and artists arranged below them.

Codes of law and religion were often used to codify these complex hierarchies. The maintenance of social order and the efficient functioning of these early states relied heavily on hierarchies, as leaders and elites were often seen as having been selected by divine decree. Furthermore, hierarchies affected how resources were managed and how surplus goods were distributed.

Challenges and Resistance

Even in the earliest human societies, hierarchy was not without its challenges and opponents.

Throughout history, there have been attempts to topple established hierarchies, as well as uprisings and social unrest. In several early societies, reformers, philosophers, and leaders advocating for more egalitarian social structures came to prominence.

The hierarchy in early human societies served a dynamic purpose, adapting to shifting cultural norms, technological advancements, and environmental changes. As societies expanded and changed, new hierarchies emerged and the relationships between power and authority continued to shift.

The evolution of human hierarchy

The study of prehistoric human societies shows how hierarchical structures progressively changed to meet the changing needs and complexities of human societies. Hierarchies evolved to manage resources, maintain order, and address the issues posed by larger and more settled populations. To comprehend the greater social development and historical trajectory of humanity, one must grasp the origins and functions of hierarchy in prehistoric human societies. It also makes clear the complex interactions between hierarchy, power, and human social dynamics that continue to shape our world today.

4.1 Hunter-gatherer groups and egalitarianism: Analyzing Ancient Social Structures

Hunter-gatherer societies, which occupied most of human history, offer an interesting framework for comprehending egalitarianism. These groups are usually viewed as more egalitarian than later, more complex societies because they led nomadic, hunting, and gathering lifestyles. Examining the connections among members of hunter-gatherer societies reveals the principles of equality that shaped their communities and the beginnings of early human social organization.

Egalitarianism in Hunter-Gatherer Communities

Resource Sharing: One of the traits that distinguish hunter-gatherer societies is the sharing of resources. Reciprocity was greatly valued in these groups, and members typically shared their food supplies. Sharing food was not only a kind gesture; it was a survival strategy. A nomadic lifestyle often involves unpredictable food availability, so sharing ensured that everyone in the group had access to food, even during periods of scarcity.

Gender Equality: In many hunter-gatherer societies, the distribution of labor between men and women was largely equitable. While men were usually responsible for hunting and women for gathering, both roles were vital to the group's existence and were not viewed as superior to one another. Women in these societies were valued for their contributions to childcare and social gatherings, and they were respected and allowed some degree of autonomy.

Small-Scale Communities: Because hunter-gatherer groups were typically made up of small bands or tribes, egalitarianism was made possible. When the population was smaller, it was easier for people to hold leaders and decision-makers accountable and to participate in the decision-making process. The high degree of interpersonal familiarity reduced the likelihood of hierarchical power imbalances.

Lack of Accumulated Wealth: In hunter-gatherer societies, there was little personal wealth or resource accumulation. Class divisions and economic disparities were avoided because wealth was shared rather than hoarded. The economic practices of these groups were therefore deeply based in equality.

Headship and Judging in Teams of Hunters and Gatherers

Hunter-gatherer societies did have hierarchical structures and decision-making processes, despite the fact that they are usually described as being fairly egalitarian. The leadership in these groups was typically situational and informal, and members were respected for their backgrounds or areas of expertise. Rather than being inherited or set in stone, leadership roles were decided by a person's ability to provide guidance or support.

Elders and Experienced People: Because of their wealth of life experience and wisdom, people in their later years were often held in high regard. Their counsel and opinions were valued when it came to making important decisions. Rather than being seen as unfailing leaders, elders were seen as sources of wisdom.

Shamans and Spiritual Leaders: In some hunter-gatherer societies, shamans and spiritual leaders had a special role in matters of religion, healing, and guidance. These individuals were typically experts in their fields and did not hold positions of leadership in the political or economic spheres.

Decision-making by consensus: In hunter-gatherer societies, important decisions were often made by consensus. Through discussions and debates, group members were able to express their opinions, and decisions were made collectively. This tactic increased a sense of shared responsibility and decreased the concentration of power.

Challenges to Egalitarianism

There were certain barriers to this ideal despite the fact that hunter-gatherer societies were generally equal. Rivalries for mates, disagreements over resources, and the sporadic arrival of charismatic leaders can upset the equilibrium. However, because these societies were smaller and more intimate, it was simpler to resolve these problems and maintain a certain degree of social equality.

The Transition of Agriculture and Complex Societies

As sedentary, agricultural lifestyles replaced hunter-gatherer societies, the social structure saw significant changes. As a result of gathering excess resources, establishing permanent settlements, and necessitating labor specialization, complex social hierarchies and disparities emerged.

Historical Perspectives

Examining hunter-gatherer societies provides significant new insights into the mechanisms of organization in prehistoric human societies as well as the emergence of egalitarianism. Gender parity, resource sharing, and informal leadership structures allowed these societies to maintain a relatively high degree of equality despite their fair share of challenges. Understanding the egalitarian principles of these early human communities can shed light on how social structures have changed over time and offer helpful advice for addressing inequality and promoting cooperation in contemporary society.

4.2 The Emergence of Social Complexity: Understanding the Path to Complex Human Societies

The rise of social complexity is one significant development in human history. It represents the transition from small, egalitarian towns to larger, more intricate communities with intricate social, political, and economic structures. An essential turning point in the development of human civilization was marked by this shift. To comprehend the components and mechanisms that resulted in the development of social complexity, we must look at the historical and archaeological evidence that illuminates this fascinating process.

The Roots of Social Complexity

Agricultural Revolution: The agricultural revolution, which began some 10,000 years ago, had a major impact on the evolution of social complexity. It marked the change from a hunter-gatherer, peripatetic lifestyle to a settled agricultural one, allowing for the steady production of surplus food. This excess provided the foundation for more complex societies and larger populations.

Food Surplus and Specialization: The abundance of food produced allowed some people to focus on pursuits other than farming. Labor specialization led to the emergence of new social classes, including farmers, artisans, and rulers. Specialization in labor contributed to increased productivity overall and the development of more sophisticated goods and technologies.

Urbanization: The concentration of surplus resources in urban areas was a significant factor in the development of social complexity. Cities evolved into major hubs for trade, politics, and culture. They provided opportunities for a variety of interactions, innovation, and the emergence of specialized professions such as clerks, administrators, and artisans.

Writing and Record-Keeping: As writing systems developed—such as Egyptian hieroglyphics and Mesopotamian cuneiform—societies were able to preserve historical accounts, judicial rulings, and cultural knowledge. Writing was essential to the consolidation of power because it gave kings the ability to impose laws and rule over larger populations.

Religion and Ideology: Religious convictions and ideologies had a big impact on how complex societies developed. Leaders often used the doctrine of divine right to defend their rule. Because they served as centers of authority, temples and other houses of worship were significant to community and resource organizations.

Centralized Governance: As societies became bigger and more intricate, structures for centralized governance emerged. Often with the aid of advisors and bureaucrats, rulers kept the peace, enforced the law, and collected taxes. An indication of social complexity was the emergence of well-organized nations and empires.

Renowned examples of complex societies

The earliest civilizations listed below are excellent illustrations of how social complexity developed:

Sumer: One of the oldest civilizations in history, Sumer emerged in southern Mesopotamia around 4500 BCE. It featured intricate irrigation systems, cuneiform writing, and temples honoring patron deities. Sumerian society is a prime example of the complex structures that arose from the abundant agricultural produce of the Tigris and Euphrates rivers.

Ancient Egypt: Known for its rise to prominence around 3100 BCE, ancient Egypt was a highly centralized society ruled by a pharaoh. The construction of colossal buildings such as the pyramids and a highly developed writing and record-keeping system serve as examples of this civilization's complexity.

The Indus Valley Civilization had planned cities, an advanced drainage and sewage system, and a weights and measures system. Around 2500 BCE, it reached its peak in what are now Pakistan and India. It is renowned for both its sophisticated urban planning and trade networks.

Ancient China: Two instances of the emergence of complex society are the Shang Dynasty *(c. 1600–1046 BCE)* and the Zhou Dynasty *(c.*

1046–256 BCE). These dynasties saw the establishment of a feudal system, the concentration of power, and the use of oracle bone script for divination and record-keeping.

Challenges and the Decline of Complex Societies

Complex societies were inherently difficult. The centralization of power, resource inequality, and external threats all played a major role in their downfall. Natural calamities such as droughts, floods, and resource depletion could potentially impact the stability of these societies.

A forward-moving voyage

The result of human invention and progress is the rise of social complexity. It represents the transition from small, egalitarian communities to larger, more organized societies capable of great achievements. Understanding the factors and processes causing this change can help one comprehend the evolution of human civilization and the struggles that have shaped our past. Social complexity continues to shape human societies' futures and has made sophisticated political systems, innovative technologies, and artistic achievements possible.

4.3 How the Agricultural Revolution and Hierarchical Changes Transformed Human Societies

The term *"Agricultural Revolution"* refers to the transition that took place in human societies from hunter-gatherer to settled agricultural practices. This enormous shift had many long-lasting consequences, one of which was the emergence of hierarchical social structures. To fully comprehend the intricate relationship between the Agricultural Revolution and the emergence of hierarchies, we must investigate the transformative processes and their effects on human societies.

An Interpretation of the Significance of the Agricultural Revolution

Although the precise beginning of the Agricultural Revolution differed depending on the region, it is generally agreed upon to have occurred approximately 10,000 years ago. It marked the change in the main subsistence techniques from hunting and gathering to crop cultivation and animal domestication. Three significant innovations were the planting of seeds, the cultivation of cereals, and the herding of livestock.

The Agricultural Revolution brought about a number of important changes, including:

Production of Surplus Food: Agricultural practices enabled the production of surplus food.

A major change from the erratic subsistence of hunter-gatherer societies was marked by this abundance. Agriculture has made food storage possible by reducing the risks associated with unpredictable resource availability.

Settlements and Urbanization: Towns were able to create permanent homes in one location as they became more and more reliant on agriculture. As a result, permanent villages were able to be established, and eventually cities and urban centers.

Specialization of Labor: If there was a consistent supply of food, some people could concentrate on non-agricultural jobs like government, weaving, metalworking, and ceramics. Specialization increased overall productivity and produced a range of skill sets.

Population Growth: Plenty of food and stable living circumstances were major factors in population growth. Larger populations could support more sophisticated social structures and technological innovations.

Complex Societies: The complexity of society increased with the development of agriculture. Centralized governance structures were developed to oversee resources, uphold the law, and collect taxes. Elites and rulers became more noticeable as hierarchical systems grew more established.

The Way Hierarchies Are Formed

The shift to agricultural economies had a major impact on the emergence of hierarchical social structures.

Several things contributed to this development:

Control of Resources: Agriculture led to the acquisition of resources and land ownership. Those in possession of agricultural surplus and fertile lands gained significant power. This control over resources served as the foundation for hierarchical structures.

Division of Labor: As a result of labor specialization, social classes were created. Administrators, artisans, and religious leaders were among the non-agricultural workers who advanced in social status. Social strata were established as a result of these statuses and positions frequently being passed down through families.

Political Centralization: As agricultural societies grew more complex, the central government was required to manage their affairs. Leaders and elites assumed leadership positions to create laws, taxation policies, and organizational hierarchies. The first empires and the rise of well-organized states are two of the best examples of power concentration.

Religion and Ideology: In many agricultural societies, religious beliefs and ideologies served to reinforce hierarchical structures. Leaders in religious organizations frequently claimed they had a divine right, and these organizations served as centers of power and control.

Well-known Agricultural Societies with a Caste System

Known as the *"cradle of civilization,"* ancient Mesopotamia was home to highly developed societies like Summer, Babylon, and Assyria. These city-states featured hierarchical priestly classes, sophisticated legal systems, and ruling elites.

Ancient Egypt: The Nile River's fertility allowed for the emergence of a hierarchical society. Egypt was the site of massive building projects and elaborate temples built during the divine rule of the pharaohs.

Indus Valley Civilization: Based on its well-planned cities like Mohenjo-Daro and Harappa, it is believed that the Indus Valley Civilization had centralized governance and social hierarchies.

China's Shang Dynasty: During this time, complex hierarchies appeared and oracle bone script, which was used for record-keeping and divination, was created.

Challenges and Modifications

Hierarchical agricultural societies were not without challenges. Power struggles, social inequality, and resource exploitation are some of the issues that can lead to social unrest. To address these problems, some societies established codified laws, bureaucratic structures, and religious doctrines.

An Era of Revolution

The Agricultural Revolution, a crucial juncture in human history, altered the fundamental makeup of societies. In the process of solving some issues, it led to the development of hierarchical social structures that brought about new complexities and inequality. Understanding this turning point provides important insights into the development of human society, its achievements, and its ongoing challenges with regard to power, inequality, and governance.

Chapter 5:

The Hierarchies of Human History: An in-depth Examination of the Evolution of Social Structures

Throughout human history, hierarchies have influenced social structures and interpersonal relationships. Studying human history reveals the evolution, significance, and variety of forms that hierarchies have taken over time. This thorough analysis of human hierarchy sheds light on the complexities of social structures and how they have affected society historically and currently.

The earliest hierarchies' chieftains and tribes

In the early ages of human history, when societies were small and reliant on familial ties, chieftainships or tribal leadership were prevalent forms of leadership. In these hierarchies, which were typically informal, leaders emerged according to qualifications such as age, experience, or track record. The primary duty of these leaders was to provide guidance and make choices when there was a dispute or a need to distribute resources. There was a hierarchy, but it was not very high, and people in the community could easily get in touch with and hold leaders responsible.

The Rise of Complex Societies: The Creation of Empires

As human societies evolved from tiny, agrarian communities to complex civilizations, hierarchical systems were born. Hierarchies became more structured as a result. The rise of empires like the Roman Empire, the Han Dynasty, and the Persian Empire led to the concentration of power under emperors or kings. These leaders held immense power and were revered as either divine or semi-divine entities. Within their hierarchical structures, these empires housed nobility, military elites, and bureaucratic systems.

These ancient empires ruled over vast swaths of territory, imposed taxes, and maintained order largely through their hierarchies. These empires built infrastructure and established social order, but they also had to maintain a hierarchical structure that often served the interests of upper class capitalists at the expense of lower class laborers.

Feudalism and the Middle Ages

In some regions of Asia and Europe during the Middle Ages, feudalism was prevalent. This hierarchical system was defined by a rigid social structure composed of kings, nobles, vassals, and serfs. Kings granted nobles land and titles in exchange for their loyalty and military service, despite their absolute power. Nobles awarded land to vassals in turn, but the majority of the population was made up of serfs who worked the land.

One essential element of feudal hierarchies was land ownership. Even though they provided stability and protection during a volatile historical period, they also limited the social mobility of the lower classes and preserved economic disparities.

The Renaissance and Enlightenment

The concepts of hierarchy and how they were contested changed in Europe during the Renaissance and Enlightenment periods. During this period of great philosophical, artistic, and cultural development, the groundwork for modern democracy was established. Thinkers like Jean-Jacques Rousseau and John Locke, who favored equality, the social compact, and individual liberty, questioned the divine right of monarchs. These ideas played a critical role in toppling long-standing hierarchies and creating more egalitarian and inclusive societies.

Modern Hierarchies and Contemporary Difficulties

Hierarchies continue to be necessary for the operation of corporate structures, the educational system, and the government even in the modern day. Modern democracies have embraced political equality and individual rights, but hierarchies persist in the workplace, in social strata, and in economic structures.

Nonetheless, there are challenges associated with these hierarchies. The main drivers of social unrest and calls for reform continue to be systemic discrimination, power imbalances, and income inequality. Movements for workers' rights, racial justice, and gender equality are a few examples of contemporary efforts to address hierarchical disparities and create more equitable systems.

The hierarchy is still changing.

The history of hierarchies in human society is one of adaptation, evolution, and change. Hierarchies have played a crucial role in maintaining social order, governance, and resource distribution, despite being a cause of inequality and social issues. The constant quest for greater equality and social justice is a reflection of the dynamic nature of hierarchies throughout human history. Understanding social issues and their evolution and impact is critical to addressing current issues and building more inclusive and equitable societies in the future.

5.1 The Rise of Empires and States: Impacting Human History

In human history, the creation of states and empires is an important and recurrent theme. This revolution has affected world politics, culture, and the development of civilizations in a significant way. Understanding the origins, procedures, and outcomes of the establishment of empires and states requires an understanding of the historical and sociopolitical contexts that have molded our world.

States and Empires: A Characteristic

Before discussing the creation of states and empires, it is important to define these terms. A king or other central authority rules over vast, multiethnic, and often multicultural political territories that are called empires. States, on the other hand, are autonomous political entities with clearly defined borders and governing bodies.

Factors Influencing the Rise of Empires and States

Military Conquest: Using force to establish empires was a common strategy. Prominent figures and military forces occupied adjacent areas in order to expand their domains of control. The empires of Genghis Khan and Alexander the Great are notable examples of conquest-driven expansion.

Trade and Economic Influence: Economic factors had a major impact on the rise of empires. Trade routes like the Silk Road and the trans-Saharan trade routes, which permitted the exchange of goods, ideas, and cultures, contributed to the rise of empires like the Roman and Ottoman.

Innovations in Governance: A small number of nations and empires developed inventive political structures. For instance, the Roman Republic contributed to the popularization of the concepts of representative government and the rule of law, which influenced the democracies that followed.

Religious and Ideological Movements: Religion and ideology played a major role in the construction of empires. Empires were established with the help of religions such as Christianity and Islam spreading throughout the world. Moreover, charismatic leaders often used ideological movements to galvanize their supporters.

Examples of Well-Known States and Empires

One of the most well-known empires in history, the Roman Empire grew throughout Europe, North Africa, and the Middle East starting in the first century BCE. Among its legacy are legal systems, imposing architecture, and languages descended from Latin.

The Byzantine Empire: The eastern half of the Roman Empire, the Byzantine Empire lasted for more than a millennium after the western half of the empire collapsed. It functioned as the center of Christian Orthodoxy, culture, and the arts.

The Ottoman Empire was founded in the fourteenth century and eventually became a major force in both Europe and the Middle East.

It was well known for its military prowess, administrative prowess, and architectural achievements.

The Chinese Dynasties: Some of the dynasties that have influenced Chinese history are the Tang, Song, Ming, and Han. These dynasties lived through periods of extensive trade networks, rapid technological development, and thriving cultures.

The British Empire was the largest empire in recorded history, spanning multiple continents and territories. Its enduring influence can be seen in the English language, legal systems, and other institutions.

Legacy of States and Empires

The rise of empires and states has had a significant impact on modern international affairs. It has impacted the growth of international relations, cultural exchange, and global economics. The influence of these empires can be seen in many aspects of contemporary society.

Empires have influenced languages, literature, art, architecture, and other aspects of culture. One example of the Roman influence in Europe is the use of languages based on Latin.

Legal Systems: Empires bequeathed enduring legal systems to their subjects. The principles of representative government, property rights, and the rule of law were established by the states and empires of antiquity.

Impact on Religion and Ideology: The activities of multiple empires played a significant role in the spread of religious ideas. The expansion of Christianity throughout the Roman Empire and the spread of Islam during the Islamic Caliphates are two instances of this influence.

Geopolitical Legacy: Present-day geopolitical borders and conflicts are often historically derived from the states and empires of the past. There are cases where the dissolution of empires has left a convoluted legacy of racial and political differences.

Challenges and Modifications

States and empires achieved tremendous progress, but they also faced challenges and frequently had to veer off course when things didn't go as planned. Among the challenges were external threats, internal strife, economic downturns, and shifts in political power.

The enduring influence of nations and empires

The creation of states and empires is a recurring theme in human history, and it has a big impact on the modern world. These political actors have had a significant impact on the design of our economies, societies, cultures, and geopolitical environment. The enduring impact of these empires highlights the significance of their rise as well as the nuanced and complex history of humanity.

5.2 Medieval Societies: An Intricate Web of Religious Hierarchies, Feudalism, and Monarchies

During the Middle Ages, feudalism, monarchy, and religious hierarchies coexisted to create a complex social structure that influenced people's lives throughout Europe and beyond. These complex systems of politics, religion, and society had a big impact on how people lived their lives and how history unfolded as a whole. To understand the dynamics of feudalism, monarchy, and religious hierarchies during the Middle Ages, we must look at the makeup, functions, and interactions of these systems.

Feudalism: A System of Ownership and Obligations

Feudalism ruled the social and economic spheres in medieval Europe, particularly in the early and high Middle Ages *(roughly the 9th to 15th centuries)*. It stood out for having a hierarchical structure based mostly on land ownership for both wealth and power. At the top of the feudal hierarchy stood the monarch, who had complete authority over all of the realm. Beneath the monarch stood the nobles, or vassals, who received land grants in exchange for their loyalty oaths and military service. This system established a web of reciprocal duties by providing lords with protection and land and providing vassals with military support.

Monarchs: The Almighty

Throughout medieval Europe, monarchies served as the main form of government. The king or queen held ultimate power and authority over their territories, often claiming a divine right. A monarch could create laws, raise armies, and give land to nobility. The nobility's relationship with the monarchy was complex because monarchs relied on their vassals for help with both military and administrative tasks. The necessity of the nobility's allegiance to the feudal system's functioning balanced the monarch's power.

Religious Hierarchies: The Influence of the Church

Strong religious hierarchies, among which the Roman Catholic Church was a major institution, were a defining feature of medieval society. The pope had enormous influence over matters of politics and religion because he was the head of the Church. The Church played a vital role in the lives of medieval Europeans by providing social services, education, and spiritual guidance. It also aided in the codification of laws and increased usage of Latin as an academic language.

Because religious leaders frequently held positions of authority and advised kings on moral and ethical matters, the Church had an impact on politics as well. The hierarchy had an additional layer of control and discipline because of the Church's authority to excommunicate groups of people or leaders. The relationship between monarchs and the Church was often characterized by a precarious balance of power, with disagreements frequently arising over issues like bishop appointments and tithe collection.

Talks and Arguments

Relationships between religious hierarchies, monarchies, and feudalism were complex and occasionally tense. Monarchs sought to increase their own authority by stifling the influence of nobles and maintaining control over the Church. On the other hand, the Church could legitimize monarchs by its support and blessings and often acted as a mediator in disputes within the realm.

Feudalism supplied the economic foundation for both monarchies and the Church, since land and agricultural productivity were necessary for both to survive. Because of the interdependence of these systems, a precarious balance was maintained in which each institution maintained a certain amount of autonomy and authority.

Barriers and Regression

Over time, the medieval world faced challenges as a result of these hierarchical structures. The growth of towns, trade, and centralized monarchies gradually threatened the feudal system. The Protestant Reformation in the sixteenth century caused religious divisions that further weakened the power of the Catholic Church.

When monarchies consolidated their power and control over territory, they often sought to weaken the Church's influence and establish their own authority over matters of faith. The complex interrelationships and tensions between these systems were eventually replaced in the early modern era by more absolutist and centralized forms of governance.

The Effects of Medieval Class Structures

The complex relationships that existed during the medieval era between feudalism, monarchies, and religious hierarchies have had a significant influence on modern societies. The concepts of authority, power, and governance that emerged during this time continue to have an impact on the political, social, and religious structures of the modern world. Analyzing these historical frameworks aids in our comprehension of the intricate relationships that have shaped human society and governance throughout history.

5.3 A Revolutionary Era: The Industrial Revolution and Contemporary Societal Structures

The Industrial Revolution, which began in the late 18th century and resulted in profound changes to the economy, technology, and social structure, is one of the most influential eras in human history. During this revolutionary era, the fundamentals of society were altered, leading to the development of modern societal structures and a significant shift in the way people lived and worked. To fully understand the intricate relationships between the Industrial Revolution and contemporary social structures, we must look at the significant innovations, developments, and impacts of this historical period.

The History of the Industrial Revolution

Following its inception in Great Britain, the Industrial Revolution swiftly spread to other nations.

It was characterized by several related developments, such as:

Mechanization: Several industries, most notably the textile industry, saw increased production efficiency as a result of mechanization. The invention of the power loom, spinning jenny, and steam engine transformed manufacturing processes.

The shift in economic power from rural to urban areas was the impetus for urbanization. People flocked to industrial centers in search of factory jobs, which fueled city growth and the development of new urban structures.

Technological Innovation: The steamship and the locomotive, two examples of transportation innovations, revolutionized trade and communication. The telegraph enabled fast information transmission over long distances.

Division of Labor: Total productivity rose as assembly line production and task specialization proliferated.

Economic Growth: The Industrial Revolution significantly influenced economic growth, particularly in capitalist economies. It led to the emergence of new markets, industries, and business opportunities.

Modern Social Groups Arise

The Industrial Revolution introduced innovations and fundamentally altered social structures, both of which continue to have an impact on modern society.

The Industrial Revolution made modern capitalism possible. As markets expanded and industrial production increased, the ideas of supply and demand, private ownership, and competition became crucial to economic systems. Capitalism remains the dominant economic system in the modern world.

Urbanization and City Structures: As people moved from rural to urban areas, modern cities and urban structures were created. The growth of cities, with their distinct social dynamics, architectural forms, and infrastructure, came to define contemporary societies.

Social Classes: Industrialization led to the emergence of new social classes. The division between the working and capitalist classes was more obvious. Labor unions and social movements that supported workers' rights and social reforms emerged as a result of the inequalities of the Industrial Revolution.

Education Systems: The demand for a highly educated workforce drove the expansion of education systems. The development of modern educational frameworks and standards was facilitated by the growth of public education.

Technological Advancements: Since the Industrial Revolution, new technologies have continued to shape and impact modern social structures. Along with a revolution in everyday life brought about by the development of electricity, telecommunication, and the internet, new communication networks and structures were established.

Globalization: The expansion of global transportation and trade networks served as the cornerstone for modern globalization. Global organizations, economies, and cultures are interdependent and have a significant impact on modern societal structures.

Challenges and Consequences

There were challenges and consequences associated with the Industrial Revolution. Rapid industrialization led to social inequality, environmental damage, and worker exploitation. It sparked labor movements and calls for regulation and reform.

The Welfare State and Social Safety Nets

In reaction to the social problems and inequality resulting from the Industrial Revolution, numerous countries instituted welfare states and social safety nets. These systems, which provide health care, unemployment benefits, and assistance for the elderly, are part of modern society structures that prioritize social well-being.

A legacy of transformation

During the period of profound transformation known as the Industrial Revolution, modern societal structures came into being. The social, technological, and economic changes of this era continue to influence the world in which we live. An understanding of the consequences of the Industrial Revolution is necessary to comprehend contemporary economic systems, urban structures, social classes, and technological advancements. It acts as a reminder of the enduring impact of this revolutionary era on modern society.

Chapter 6:

Analyzing the Complex Networks of Human Behavior: The Psychological Basis

Human psychology is the study of the many and varied aspects of the human mind and behavior. To understand the psychological underpinnings of human behavior, one must delve into the depths of motivation, emotion, and cognition. The nuances that affect our attitudes, relationships, and behaviors are made clearer by this investigation.

Cognition, the Mental Engine of Thought

Cognition is the general term for the mental processes that underlie human knowledge and thought. There is a need for perception, concentration, memory, language, problem-solving, judgment, and reasoning. Understanding how people acquire, organize, process, and apply information is the aim of cognition research.

Perception: Perception is the process through which sensory data are transformed into meaningful experiences. It comprises the interpretation of auditory, tactile, and visual data obtained from the external environment. A few things that influence perception are memory, attention, and past experiences.

Memory: Memory is the ability to store and retrieve information. Three different types of memory comprise it: sensory, short-term, and long-term. Our memory and cognitive functions are greatly influenced by the encoding, storing, and retrieval processes.

Language: Language is a crucial component of cognition. It helps with communication as well as the expression of ideas and thoughts. Language study examines language structure, learning, and processing in great detail and demonstrates how language impacts our mental and social processes.

Making decisions and solving problems are everyday tasks for people. These processes require locating and evaluating options in order to reach a decision or find a solution. The emotional and cognitive factors that influence our decisions are examined in the study of problem-solving and decision-making.

Sensations: A Bright Color Palette of Emotion

Emotion is a basic aspect of the human experience. Emotions are intricate psychological and physiological responses to environmental and internal stimuli. They have a huge impact on our behavior and interpersonal relationships. Emotions encompass a wide range of feelings, from joy and love to fear and rage.

Theories of Emotions: Several theories have been put forth in an attempt to explain the nature and genesis of emotions. The James-Lange theory states that physiological responses come before emotional experiences.

However, the Cannon-Bard theory contends that emotions and physical responses occur simultaneously. The Schachter-Singer two-factor theory, which integrates physiological responses and cognitive appraisal, generates emotional experiences.

Emotion regulation is the ability to manage and regulate one's emotional responses. Social interactions and mental health both depend on emotional regulation. Techniques for controlling emotions include cognitive reappraisal, diversion, and suppression.

Motivation: The Inherent Force of Action

Motivation is the fuel that drives human behavior. This process initiates, directs, and maintains goal-directed actions. Understanding motivation requires looking into the factors that influence behavior, such as needs, desires, rewards, and goals.

There are two types of motivation: intrinsic motivation, which originates from internal desires or interests, and extrinsic motivation, which is derived from external rewards or pressures. Achieving goals and sustaining motivation necessitates finding a balance between these driving forces.

Maslow's Hierarchy of Needs: Abraham Maslow created a theory of human motivation using a hierarchical framework. The necessities for survival **(food, shelter, etc.)** are the most fundamental needs, followed by those for safety, respect, love and belonging, and self-actualization. This theory highlights the ways in which different needs affect people's behavior.

Self-Determination Theory: This theory maintains that relatedness, competence, and autonomy are essential human psychological needs. These needs have to be satisfied for intrinsic motivation and well-being.

Social and Cultural Influences: Motivation is also influenced by social and cultural factors. Social expectations, peer pressure, and cultural norms all have an impact on an individual's behavior and aspirations.

The Relationships Among Psychological Foundations

Cognition, emotion, and motivation are all interconnected. Thoughts can influence emotions, and emotions can influence motivation and decision-making. One may be motivated to seek safety or avoid a threat, for example, by the emotion of fear brought on by a cognitive perception of a threat.

Understanding the psychological underpinnings of human behavior requires embracing an integrated perspective that acknowledges the interplay of these components. By exploring the complex and dynamic domain of human psychology and behavior through this examination of cognition, emotion, and motivation, we can gain a deeper understanding of both ourselves and those around us.

6.1 The Human Brain and Social Hierarchies: Untangling the Neuroscience of Social Structure

Social hierarchy is one of the fundamental elements of human civilization. Individuals are usually found in small groups or large organizations within organized social orders.

The human brain is crucial for the creation and negotiation of these hierarchies due to its complexity and adaptability. This study delves into the neuroscience of social structure and sheds light on the underlying mechanisms driving social hierarchies and the human brain.

Social Hierarchies: A Universal Human Experience

All societies and eras have social hierarchies. Within a group or community, these hierarchies categorize individuals based on their various roles, responsibilities, and social standings. They could be based on factors such as age, gender, wealth, occupation, or social skills. Hierarchies are helpful for distributing resources and assigning authority, but they also encourage cooperation.

The Brain's Social Circuitry

The human brain processes and reacts to social hierarchies in a multifaceted manner. Social cognition is the process of perceiving, understanding, and responding to social information. It involves numerous brain regions and neural circuits.

These brain regions comprise, among other things:

Prefrontal Cortex: The prefrontal cortex, particularly the medial prefrontal cortex, is a crucial part of social cognition. It helps with empathy, decision-making in social contexts, and understanding the mental states of others.

Amygdala: This area of the brain is in charge of processing emotional and social information. It is crucial for recognizing and responding to emotional cues in other people, as these affect opinions and social interactions.

Anterior Cingulate Cortex: Pain and social isolation are associated with this region. One source of distress for an individual is feeling left out or excluded in a social hierarchy.

Mirror Neurons: Mirror neurons are thought to act as a mediating factor in empathy and the ability to understand and replicate the actions and emotions of others. They support the social cohesiveness that occurs within hierarchical structures.

Position and the Sense of Hierarchy

The human brain is highly attuned to how social status is perceived within hierarchies. The brain automatically deduces status cues from nonverbal indicators like body language, tone of voice, and facial expressions. Research has shown that people can form opinions about their status quickly and that these opinions affect feelings, choices, and social interactions.

People may show signs of increased awareness and sensitivity to the social hierarchy, such as increased brain activity, when interacting with someone of a higher status. However, experiencing a sense of inferiority can trigger neural responses associated with social stress.

The Impact of Social Hierarchies on the Brain

Social hierarchies have a profound effect on brain function and structure.

As an illustration:

Stress Responses: People at the bottom of the hierarchy typically experience higher levels of stress, which modifies the physiology of the brain. Prolonged stress can affect the structure of the hippocampal region, which is linked to learning and memory.

Identity and Self-Perception: A person's position within a social hierarchy can have an impact on their sense of self. People with negative self-concepts and low self-esteem may feel inferior to others, which can have an impact on how their brains work.

Social Learning and Adaptation: Due to the neuroplasticity of the brain, individuals are able to learn and adapt to their position in a hierarchy. This adaptability is demonstrated by the development of social skills, emotional regulation, and strategic decision-making.

The Roles of Oxytocin and Dopamine

Dopamine and oxytocin are two neurotransmitters that play a role in social hierarchies. Whereas oxytocin is associated with social bonding and trust, dopamine is linked to motivation and rewards. These neurotransmitters have an impact on social behavior, motivation, and the formation of alliances within hierarchies.

Challenges and Ethical Concerns

Social hierarchies can be beneficial as well as detrimental. Positive traits include collaboration, task delegation, and focused efforts toward common goals. Conversely, drawbacks could involve discrimination, exclusion, and power imbalances.

Moral questions arise from a deeper understanding of the neurology of social hierarchies. Applying this knowledge responsibly is essential, addressing issues of equity, justice, and the welfare of individuals in hierarchies.

The intricate connection

The human brain and social hierarchies have a dynamic and intricate relationship. Understanding how people navigate and adapt to social structures requires an understanding of social cognition, status perception, and response to cues related to hierarchy. This study of the neuroscience of social structure advances our knowledge of the connection between societal structure and human psychology by shedding light on the complex web of human interaction and behavior within hierarchies.

6.2 Social Hierarchy: The Dynamics of Dominance, Power, and Status

Status, power, and social dominance are the fundamental building blocks of human societies. These concepts form the foundation of the hierarchies that affect our relationships, interactions, and institutions. Understanding the dynamics of status, power, and social dominance is essential to comprehending the complexities of human social systems and the factors that shape our decisions and behaviors.

Status: Position in the Hierarchy of Command

A person's position or standing in a social hierarchy is referred to as their *"status"*. It is an expression of how important, powerful, and socially significant they are in that specific context. A person's status can be determined by their accomplishments, such as a successful education or career, or it can be inherited based on traits like age, gender, or ancestry.

Two types of status: assigned and acquired

Ascribed Status: A person's ascribed status is determined by their innate characteristics, such as gender, race, or ethnicity, which can be inherited from birth or developed over time. Social hierarchies are frequently significantly impacted by these traits.

Achieved Status: On the other hand, achieved status is the result of personal effort and accomplishments. One's educational background, professional achievements, and personal traits all play a role in their attained status. It conveys the idea that someone's skills, knowledge, or contributions are appreciated.

Power: The Ability to Influence and Direct

Power is the ability of an individual to control and direct resources or other people in a social hierarchy. It can manifest itself in a variety of ways, including as physical, political, or economic power. There are two sorts of power: formal power, such as political leadership, and informal power, such as the influence of a community leader.

Power comes from coercive, rewarding, expert, and referent sources.

The ability to exert control over others through the threat or actual use of force is known as coercive power. There may be physical violence or the threat of serious harm.

Reward Power: The basis of reward power is the ability to persuade others by providing incentives or rewards. Exchange systems and economic transactions make this clear.

Expert Power: Expert power comes from possessing particular skills or knowledge that are respected by others. People often respect those who are experts in their fields.

Referent Power: The core elements of referent power are identification, respect, and interpersonal attraction to an individual. It's the kind of power that comes from being liked, respected, or esteemed.

Social Dominance: The Struggle for Control

Social dominance is the exercise of control, power, and authority within social hierarchies. It is the external expression of rivalry and power dynamics over status, influence, and resources. Social dominance can be demonstrated in a variety of ways, from subtle deeds to overt claims of power.

Challenges and Discussions

The dynamics of status, power, and social dominance are not without challenges and conflicts.

These include:

Inequality: Uneven distribution of status and power can lead to social inequality, in which some individuals or groups gain disproportionately while others experience unfavorable outcomes.

Abuse of Power: When power is misused, oppression, authoritarianism, and human rights abuses can result. Ensuring the ethical and responsible use of power is essential.

Evolution and Resistance: Social structures are always changing. They can be challenged and altered by social movements, activism, and political change.

The Interface of Status and Power

The interplay of status and power creates a complex web of social dynamics. High status individuals might have easier access to resources, opportunities, and education, all of which could strengthen their position. However, people in positions of power have the power to shape and influence how society perceives status, giving greater weight or distinction to certain traits or achievements.

Understanding the hierarchy of people

Due to social dominance, status, and power, human hierarchies affect our interactions, opportunities, and life experiences. The evolution of our political, economic, and social structures is based on these concepts. To address social injustices, cultivate moral leadership, and create more just and equitable societies, it is imperative to understand the dynamics of status, power, and social dominance. It offers comprehension of the subtle relationships that exist between behavior, motivation, and decision-making in people within the intricate web of social structures.

6.3 The Pull Toward Social Equity: Collaboration and Aversion to Inequality

Cooperation and inequality aversion are two significant and closely related facets of human social behavior. Understanding these dynamics is essential to understanding the complexities of human interactions, especially in social, political, and economic contexts. This study examines the intricate connection between cooperation and inequality aversion, shedding light on how social norms, justice, and empathy affect cooperative efforts.

Avoiding Inequality and Seeking Fair Treatment

The phrase *"inequality aversion"* refers to the discomfort or desire to minimize the unequal allocation of advantages or resources. It stems from the fundamental human sense of equity and justice. Behavioral economics and psychology research shows that people often exhibit an aversion to inequality, preferring more equitable distributions of resources or outcomes.

Ultimatum Game: Showing inequality aversion, the ultimatum game is one of the oldest experimental paradigms. A player known as the proposer receives a set sum of money in this game, and they have to give some of it to a player known as the responder. The respondent may choose to accept the offer and move forward with the suggested split, or it may choose to reject it and forgo any further payment to either player. Even though it would be economically rational for the respondent to accept any offer, people frequently decline offers that they feel are unfair. This behavior suggests a distaste for unequal distribution.

Working Together: The Cornerstone of Social Engagement

Cooperation is the act of individuals working together to achieve common goals or find solutions to problems. It is a fundamental aspect of human civilization that underpins a wide range of activities in daily life, such as politics, economics, and interpersonal interactions.

Prisoner's Dilemma: The prisoner's dilemma is a well-known game in game theory that highlights the tension between cooperation and self-interest. In this game, two players can choose to cooperate or betray their partner. While mutual benefit dictates cooperation, self-interest can lead to betrayal. The prisoner's dilemma emphasizes the need for cooperation for shared benefit as well as the challenge of balancing individual and group interests.

The Connection Between Inequality Aversion and Cooperation

Intimate relationships exist between cooperation and inequality aversion, and these relationships have a significant impact on behavior.

Promotion of Fairness: Individuals who are opposed to inequality might find motivation to cooperate in order to bridge divides and search for more equitable allocations. This natural sense of justice frequently inspires cooperation among team members to achieve just outcomes.

Reciprocity norms: The desire for justice and reciprocity is a fundamental aspect of cooperation. When people work together, they hope to be treated well in return. People are more likely to cooperate when they perceive others as fair and trustworthy, which is why reciprocity norms promote cooperative behavior.

Collaboration and Collective Action: To achieve common goals, collaboration is essential in circumstances that call for collective action, such as preserving the environment or providing public goods. Inequality aversion may serve as an incentive for participation in these cooperative efforts to solve shared problems.

Impact of Social Norms: The degree to which cooperation and inequality aversion are balanced is greatly influenced by cultural values and social norms. Societies that prioritize fairness and impartiality might also exhibit greater collaboration.

Challenges and Discussions

The relationship between inequality aversion and cooperation is not without challenges:

Cultural Variations: Cultural norms and values can be used to explain variations in the level of cooperation and aversion to inequality. Cooperation levels can be impacted by disparate cultural conceptions of justice.

Studies have uncovered a phenomenon called the *"inequity aversion paradox,"* which states that individuals who strongly object to inequality may choose not to cooperate when doing so could lead to unequal distributions.

In some cases, people use inequality aversion as a tactic to manipulate or control cooperative relationships in order to achieve their own goals. This is known as strategic behavior.

The Act of Equilibrium

Human social behavior is characterized by aspects of cooperation and inequality aversion. Although people's dislike of inequality drives them to seek justice and reduce inequality, cooperation is the method by which group goals are achieved. Social norms, cultural values, and the context of the interactions all influence how these forces interact. Finding a balance between one's own interests and the pursuit of equity is still a challenging and dynamic task in human societies, with profound implications for political, social, and economic dynamics. Understanding this interaction improves our comprehension of social justice, interpersonal cooperation, and the intricate web of human relationships.

Part III:

Modern Views

Chapter 7:

Managing Authority and Influence in the Virtual World: Digital Age Hierarchies

As the digital age has grown, communication and information exchange have entered a completely new era. In this rapidly evolving landscape, hierarchies take on unique dimensions and forms that distinguish them from traditional structures. Understanding hierarchies in the digital era is necessary to comprehend how cooperation, influence, and power manifest in the virtual world. This study examines the complexities of digital hierarchies and how they impact our day-to-day lives.

Digital Enterprises: The Revolution

Also known as online hierarchies or virtual hierarchies, digital hierarchies are power and influence structures that exist in the online sphere. They can be found in a variety of settings, including traditional organizational hierarchies found in businesses and unofficial social media networks. Numerous factors, such as algorithms, follower counts, user engagement, and expertise, have an impact on digital hierarchies.

Crucial Elements of Digital Organizations

In the digital age, visibility and exposure play a critical role in determining an individual's position in a hierarchy. Celebrities and influencers are two instances of extremely visible users who regularly hold prominent positions in online hierarchies.

Knowledge and Expertise: Expertise in a particular field is highly valued in digital hierarchies. People who can provide incisive analysis, helpful information, or solutions usually gain respect and notoriety.

Follower Counts and Social Capital: The number of followers, subscribers, or connections an individual has on social media platforms is a commonly used metric to assess their influence. Users with large followings usually occupy higher positions in digital hierarchies.

Algorithms and Recommendations: Social media sites and search engines use algorithms that have an impact on user and content visibility. Because they control the content that users see, they are crucial to the formation of digital hierarchies.

Digital Companies in Operation

Digital hierarchies are prevalent in a lot of online settings.

Social media: The three primary elements that establish the hierarchies on social media platforms like Facebook, Instagram, and Twitter are user engagement, follower counts, and influencer status. Users with a sizable fan base have the ability to shape public opinion and trends.

Online Communities: On Internet forums, subreddits, and community-based platforms, hierarchies are determined by a variety of factors, including reputation, contributions, and the ability to manage content.

E-commerce and marketplaces: Online marketplaces such as Amazon and eBay have hierarchies based on product rankings, seller ratings, and user reviews. These hierarchies influence consumer choice.

Digital Organizations: In the digital age, businesses and institutions may create virtual hierarchies to support remote work and online collaboration. These hierarchies might not be the same as conventional office setups.

Challenges and Intricacy

Digital hierarchies come with their own set of challenges.

Algorithmic Biases: Digital platforms' algorithms may inadvertently reinforce biases and inequality, which can have an impact on how digital hierarchies are organized.

Cyberbullying and harassment: The anonymity and distance provided by the digital sphere can lead to instances of cyberbullying and harassment, which can upset online hierarchies.

Privacy Concerns: Digital hierarchies and surveillance raise concerns due to the collection of personal information and the potential for privacy violations.

Information manipulation: The spread of misinformation, fake news, and false information has the ability to change digital hierarchies and public opinion.

The Impact on Humanity

Understanding digital hierarchies is crucial because they have a significant impact on society. They affect consumer choices, public opinion, information exchange, and even political outcomes. Recognizing the ways in which digital platforms and algorithms create hierarchies is essential to addressing issues with misinformation spreading, echo chambers, and filter bubbles.

Navigating the Digital Labyrinth

Hierarchies are intricate, dynamic, and fundamental to the modern human experience in the digital age. As we continue to navigate this dynamically shifting digital landscape, it is imperative that we comprehend the power dynamics and structures that shape our online interactions. As users, producers of content, or digital citizens, we can all engage with the virtual world responsibly and thoughtfully when we comprehend digital hierarchies. Ultimately, this aids in molding the digital era into a space that embodies our common ideals and objectives.

7.1 Digital Social Environment Navigation: Social Media, Virtual Communities, and Power

The advent of social media and the internet has fundamentally transformed the ways in which we interact, exchange knowledge, and influence each other. Social media and online communities have given rise to new power dynamics and communication channels that are having an impact on global society. Understanding the complex relationships between social media, online communities, and influence in today's world requires an understanding of the dynamics of these digital spaces.

Social Media: Amplification of Voices

Social media platforms, including Facebook, Instagram, Twitter, LinkedIn, and others, have grown into powerful instruments for communication, influence, and self-expression. These platforms provide individuals, companies, and influencers with access to diverse audiences, igniting debates, spreading knowledge, and swaying public opinion.

Important Influencers on Social Media:

Engagement and Follower Counts: A large number of subscribers or followers on social media is frequently used as a gauge of influence. Users with a large following can impactfully reach a larger audience through their content. Likes, comments, shares, and other forms of high interaction also have an impact on influence.

Influencer Marketing: As social media influencers grew in number, so did influencer marketing. Influencers and brands collaborate to promote products, ideas, and services to their passionate and engaged fan communities.

Trending Topics and Virality: Hashtags, subjects, and content can become viral on social media very quickly. Significant posts and trends have the ability to captivate people worldwide and spark discussions.

Algorithmic Influence: On social media platforms, algorithms are used to select which content appears in users' feeds. Understanding these algorithms and making content visible are key components of influence.

Online Communities: The Potential of Niche Networks

Online communities—found on sites like Reddit, forums, and specialized social networks—offer a unique form of influence. These communities provide spaces for people to interact, exchange information, and influence one another. They are often based on shared passions, identities, or interests.

Crucial Components of the Impact of Online Communities:

Reputation and Expertise: In virtual communities, individuals possessing a strong online persona, an abundance of valuable insights, and proof of their expertise can significantly influence their particular niche.

Moderation and Governance: In some online communities, leaders or moderators keep an eye on conversations and impose guidelines. These individuals have the power to influence the direction and tone of the community.

Niche Expertise: Online communities often foster the development of niche expertise. Proficient individuals in a specific domain or topic have the ability to impact others by sharing their expertise and insight.

Online communities can unite to take collective action, including advocating for laws, raising awareness of specific issues, or advancing causes. These communities' ability to inspire a passionate user base makes them extremely powerful.

Challenges and Ethical Concerns

There are issues and ethical conundrums with social media dynamics, online communities, and influence as well.

Echo Chambers: Due to social media algorithms and online communities' self-selection, which limits the diversity of ideas, people who are exposed to information and opinions that confirm their preexisting beliefs may find themselves in these settings.

Misinformation and Disinformation: Due to the rapid spread of false information on social media platforms, the public may grow cynical and mistrustful.

Privacy Concerns: Because of the collection of user data and potential for privacy violations, these platforms' data security and monitoring procedures have come under examination.

Mental Health: The often idealized and curative representations of other people's lives on social media can be detrimental to mental health, exacerbating issues such as anxiety and social comparison.

The Impact on Humanity

Online communities, social media, and influence are vital to many aspects of modern society.

Political Influence: During elections, social media platforms have been instrumental in shaping political discourse and voter turnout.

Social Movements: Online communities have contributed to the growth of movements such as Black Lives Matter, the #MeToo movement, and climate activism.

Business and Marketing: Influencer marketing on social media has completely transformed the advertising landscape as brands try to capitalize on the popularity of content creators.

Global Connectivity: These digital spaces have enabled cross-border collaboration and cross-cultural exchange by bringing individuals and communities virtually together.

A New Impact Era

The digital age has ushered in a new era of influence, where individuals, groups, and organizations can use online communities and social media to shape public opinion, advocate for change, and connect with like-minded people. To successfully navigate this ever-changing landscape, one must possess a critical understanding of the platforms, algorithms, and moral dilemmas that underpin digital influence. As we continue to discuss the impact of social media, online communities, and influence on society, it is imperative that we strike a balance between the freedom of expression and responsible digital citizenship. Ultimately, this will contribute to molding the digital era's impact in a manner that aligns with our common goals and principles.

7.2 The Influence of Technology on Hierarchies: From the Industrial Revolution to the Digital Age

The complex and dynamic relationship between technology and hierarchies is intimately related to the advancement of human civilization. From the workplace and society to information sharing and global influence, technology has always been essential to the creation and upkeep of hierarchies in a number of contexts. The complex effects of technology on the formation and development of hierarchies are examined in this investigation.

Historical Context: Technology and the Industrial Revolution

The beginning of the Industrial Revolution in the late 1700s marked a significant change in the way that technology influenced hierarchies. The economy and social structures have undergone radical transformations due to technological advancements in manufacturing, transportation, and communication. The factory system, which was powered by machinery and automation, established a new workplace hierarchy where managers and factory owners had significant control over employees.

Using Technology in the Workplace

Technological innovations have a big impact on workplace hierarchies:

Automation and Employment Displacement: Automation technologies have caused changes in the employment hierarchy and the displacement of jobs, ranging from the steam engine to artificial intelligence. While these technologies have increased productivity, they have also disrupted traditional job roles and created new hierarchies based on skill sets related to technology management.

Knowledge-Based Hierarchy: As humankind has matured and become more sophisticated, knowledge and skill have taken precedence in hierarchies. Individuals possessing specialized knowledge in fields such as software development, data analysis, and digital marketing are increasingly valued in the workplace.

Digital Platforms and the Gig Economy: The gig economy, which has been stimulated by the expansion of digital platforms, allows people to work independently or as freelancers. In this instance, internet reputation, ratings, and client endorsements can affect hierarchies.

Communication and Information Exchange

Thanks to technology, the media and communication sectors have witnessed a revolution in the establishment of hierarchies and the sharing of information:

Media Influence and Gatekeepers: In traditional media, gatekeepers controlled the flow of information. People can now access information more easily and have access to new avenues for media influence thanks to the digital age, many of which are powered by algorithmic hierarchies that regulate content visibility.

Social media platforms have facilitated the emergence of microcelebrities and influencers who possess the capacity to establish personal fan hierarchies and exert a worldwide influence on public discourse.

Misinformation and disinformation: The spread of misinformation has been facilitated by technology, upending long-standing norms of legitimacy and veracity in the information environment.

Global Influence and Geopolitics

Technology has changed the global influence and geopolitical hierarchies:

Digital Superpowers: Often referred to as "digital superpowers," these are the nations with the greatest levels of innovation and technological advancement that have had a significant influence on world affairs. These countries set global norms and have a big say in how the world is governed.

Cyber Warfare and Security: The rise of cyberwarfare has resulted in a new stratum of authority in the domains of national security and defense.

Global Networks: As a result of technological interconnectedness, which has challenged established power structures and diplomatic hierarchies, global networks and organizations have grown.

Challenges and Ethical Concerns

There are challenges associated with technology's influence on hierarchies:

Digital Divide: Access to technology and digital resources, particularly in the areas of education and job opportunities, can lead to inequalities and hierarchies.

Privacy and Surveillance: The use of technology for data collection and surveillance raises questions about privacy and control over personal information.

Ethical Concerns: Because technology plays a part in creating or upholding hierarchies, such as in AI and algorithmic bias, fairness, discrimination, and social justice have emerged as ethical issues.

A shifting landscape

Technology has had a significant impact on political, social, and economic systems as its influence on hierarchies has grown and changed over time. It has upended long-standing hierarchies in some domains, but it has also created new challenges and hierarchies in others. To navigate this ever-changing environment, one must possess a sophisticated understanding of the ways in which power, technology, and society intersect. We must address the moral questions this raises, promote diversity, and use technology to advance society as a whole as we continue to adapt to the ever-changing effects of technology on hierarchies.

7.3 Virtual Organizations and Distributed Leadership: Managing the Future of Work

The concepts of distributed leadership and virtual organizations have emerged as a result of the introduction of technology and the changing nature of the modern workforce. In the digital age, work is becoming more and more decentralized, with teams working virtually and distributed across borders. In the context of virtual organizations, understanding distributed leadership principles is crucial for optimizing efficiency and effectiveness in this new work environment.

Virtual Businesses: Revolutionizing the Workplace

Virtual businesses are not the same as conventional brick and mortar establishments. They are also known as geographically dispersed or remote organizations. Workers in virtual organizations work remotely, often from home, collaborating and completing tasks with digital tools and communication technologies. These businesses function on the principles of flexibility, adaptability, and fruitful virtual collaboration.

Key Characteristics of Virtual Businesses:

Remote Workforce: One common feature of virtual organizations is a workforce that operates remotely from different locations. This model promotes work-life balance and employee flexibility.

Digital Tools: Digital technology is the cornerstone of virtual organizations. Tools like video conferencing, project management software, and cloud-based storage enable remote collaboration.

Geographic Diversity: Teams in virtual organizations may have members who are spread across different time zones, continents, or geographical areas. There are benefits and drawbacks to this variety.

Reduced Overhead: Virtual businesses typically have lower overhead costs than traditional offices because they don't require real office space or related expenses.

Dispersed Leadership: A Novel Strategy

Allocating leadership responsibilities to multiple individuals within an organization rather than centralizing them in a single hierarchical structure is the idea behind distributed leadership. Because distributed leadership enables leaders at various levels to make decisions and effectively manage their teams, it takes on new significance in virtual organizations.

Principles of Distributed Leadership for Virtual Teams:

Empowerment and Autonomy: Decentralized leadership fosters the capacity of team members to assume accountability for their tasks and make independent decisions. In virtual environments, where there might not be as much direct supervision, this independence is extremely helpful.

Collaborative Decision-Making: To make decisions, leaders in virtual organizations typically consult with their teams. This tactic leverages the diverse experiences and perspectives of remote team members.

Open Communication: Effective communication is essential in virtual organizations. Distributed leaders place a high importance on open and transparent channels of communication to ensure that everyone is in agreement.

Adaptive Leadership Styles: Leaders in virtual organizations may need to change their style of leadership in order to meet the unique needs and challenges encountered by remote teams. This adaptability ensures that leadership stays effective and responsive.

Challenges and Strategies

Challenges specific to virtual organizations and distributed leadership exist.

Problems with Communication: It can be challenging for remote teams to keep up effective communication. Techniques like regular video conferences, succinct writing, and the use of collaboration tools can help close communication gaps.

Team Cohesion: In virtual organizations, fostering a sense of purpose and unity among team members requires deliberate effort. Activities for team building, virtual social gatherings, and recognition programs can all contribute to a person's sense of belonging.

Evaluation of Performance: It can be challenging to evaluate the performance of a remote team member. Setting clear expectations, giving frequent feedback, and utilizing key performance indicators ***(KPIs)*** are some strategies to address this challenge.

Cybersecurity and Data Protection: Virtual organizations must prioritize cybersecurity and data protection in order to safeguard sensitive data. Strong security measures must be implemented, and team members must receive best practices training.

The Future of Work's Hybrid Model

The concepts of virtual organizations and distributed leadership will likely play a major role in the workplace of the future. A growing number of companies are introducing hybrid work models, which combine remote and on-site work, to give employees more flexibility. As the lines between traditional and virtual organizations become increasingly blurred, one's ability to successfully implement distributed leadership in a variety of work settings will depend on that ability.

Closing Remarks: Embracing the Digital Revolution

Virtual organizations and distributed leadership have profoundly altered the way we operate and lead in the digital age. To take advantage of virtual work arrangements, organizations and their leaders must adapt to these changes as technology advances. In light of the evolving nature of the modern workforce, virtual organizations can thrive by promoting autonomy, adopting adaptive leadership, and cultivating teamwork. This will guarantee success and productivity in a world that is becoming more virtual.

Chapter 8:

Issues that Cut Across Justice, Sustainability, and Inequality in the Modern World

Currently facing the global community are three interrelated challenges: sustainability, justice, and inequality. These problems are inextricably linked because inequality in wealth, opportunities, and access to resources frequently crosses paths with justice issues and has a substantial impact on the sustainability of our planet and societies. A comprehensive approach to addressing these pressing issues requires an understanding of the complex relationships that exist between justice, sustainability, and inequality.

Inequality Is the Root of Social and Environmental Divides

Inequality encompasses differences in general well-being, wealth, income, education, and access to healthcare. It is a global issue that affects individuals, groups, and nations everywhere in the world. Specifically, the ease with which individuals can access opportunities and resources is impacted by economic inequality. Injustices in society and the environment are also commonly the outcome.

Environmental Inequality: A subset of inequality, environmental injustice occurs when marginalized communities often bear the brunt of environmental degradation and climate change. These communities typically lack the resources necessary to adapt to environmental challenges, making them more vulnerable to the harmful effects of pollution, resource depletion, and natural disasters.

Economic inequality: Disparities in wealth and income result in unequal access to opportunities for work, education, and healthcare. An excessive amount of economic inequality can limit social progress and the resources needed for a sustainable way of life.

Equity: Completing the Gap

Justice in the context of inequality is ensuring that all people have equal access to resources, opportunities, and a reasonable standard of living. To achieve justice, it is necessary to address social and environmental injustices, economic inequality, and systemic discrimination.

Social Justice: Social justice advocates for treating individuals and groups equally, without regard to factors such as race, gender, or socioeconomic status. In order to promote greater equity in society, it is imperative to address social injustices such as discrimination based on race and gender.

Environmental Justice: Addressing how marginalized communities are disproportionately impacted by environmental degradation is the aim of environmental justice. Dealing with issues like pollution, the availability of clean water, and how climate change impacts already vulnerable communities are all part of this.

Sustainability: Maintaining Equilibrium Between the Past and Future

Sustainability is the ability to meet present needs without compromising the ability of future generations to meet their own needs. Sustainability requires minimizing negative environmental effects, granting everyone equitable access to resources, and engaging in responsible resource management.

Environmental Sustainability: Preserving ecosystems, halting climate change, and protecting natural resources are three crucial aspects of environmental sustainability. Making sure that resources are used in a balanced manner and cutting waste are two crucial rules.

Social Sustainability: Social sustainability includes addressing issues of inequality and poverty as well as fostering inclusive and healthy communities, education, and general well-being. Creating societies where people can prosper is essential.

Building robust, resilient economic systems that can satisfy both the needs of the present and the future is the aim of economic sustainability. Sustainable economic models aim to promote equitable wealth distribution and reduce inequality.

Overcoming Issues and Remedies

In order to address the intersection of justice, sustainability, and inequality, numerous strategies are required:

Policy and Governance: Governments and international organizations can promote social justice, environmental sustainability, and economic equality through the implementation of policies. This includes initiatives to close the income gap, protect vulnerable communities, and promote eco-friendly lifestyles.

Education and Awareness: Raising public awareness of the detrimental impacts of injustice, inequality, and unsustainable behavior is essential. An educated person is better able to advocate for change and make informed decisions.

It is crucial to support marginalized communities in participating in decision-making processes, especially when it comes to matters that affect the environment and their well-being. We call this kind of decision-making inclusive.

Economic Reforms: Restructuring economic systems to impose fair taxes, reduce wealth disparities, and promote socially responsible corporate practices can help create a more sustainable and equitable world.

Sustainable Development Goals (SDGs): The UN's Sustainable Development Goals offer a global framework for addressing these interconnected problems. The goals address issues of inequality, justice, and sustainability and provide a way forward for collective action.

A cohesive approach

Inequality, justice, and sustainability are the three sides of a complex and interconnected triangle. Ill-thought-out solutions can arise from concentrating on one issue without considering the others. A comprehensive strategy that recognizes the intersections and interdependencies of these issues is necessary to create a more just, equitable, and sustainable world. It requires coordinated international efforts, cross-sector collaboration, and a commitment to ensuring that the wellbeing of all people and the health of the planet are given equal priority.

8.1 Analyzing the Relationship between Economic Inequality and Social Hierarchies

In human society, the distribution of opportunities, power, and resources is influenced by social hierarchies and economic inequality. Differences in wealth and income are referred to as economic inequality, while social hierarchies refer to the more general social stratification structures that include factors like occupation, gender, race, and education. Understanding the complex relationship between economic inequality and social hierarchies is essential to understanding modern societies and addressing issues of social justice and equity.

Economic Inequalities: The Divide in Income

Economic inequality is the term used to describe the unequal distribution of wealth and income within a society or between nations. The income distribution is commonly quantified using metrics such as the Gini coefficient. Economic inequality can be caused by a variety of factors, including differences in income, access to education, economic policies, and the accumulation of wealth.

The causes of economic inequality include:

Wage Disparities: Pay disparities based on occupation, experience level, and education aggravate income inequality. Workers with higher levels of education and skill typically earn more money than those with lower levels of education and skill.

Wealth Accumulation: Variations in the way wealth is accumulated also play a role in economic inequality. People who own assets—like real estate, investments, and inheritances—usually have better financial standing.

Economic Policies: The distribution of income can be influenced by tax, labor, and government assistance programs. Two strategies to lessen economic inequality are progressive taxation and social safety nets.

Levels of Stratification in Social Hierarchies

The multi-level structures known as social hierarchies determine a person's status and position within a community. In addition to economic factors, these hierarchies may also take into account factors like age, gender, race, and education.

Types of Social Hierarchies:

Racial and Ethnic Hierarchies: Racism and discrimination can lead to differences in healthcare, employment opportunities, and educational opportunities, all of which worsen economic inequality.

Gender Hierarchies: Opportunities and income still differ based on gender in many societies. The underrepresentation of women in leadership roles and the gender pay gap are two examples of how gender-related economic inequality presents itself.

Educational Hierarchies: Differences in access to high-quality education and educational attainment can contribute to social hierarchies. Those with advanced degrees typically have higher earning potential.

Occupational Hierarchies: Occupational classification is often based on the abilities and credentials required for a particular position. Paying jobs typically require specific educational backgrounds and skill sets.

The Interaction: How Economic Inequality Shapes Social Hierarchies

Economic inequality affects and reinforces social hierarchies in a variety of ways:

Access to Opportunities: People with higher incomes and greater wealth have easier access to opportunities such as networking, excellent healthcare, and education. This access could preserve current social hierarchies.

Power and Influence: Economic resources are often linked to political power and influence.

Richer individuals and groups have the ability to influence legislation that could either strengthen or weaken social hierarchies.

Reproduction of Advantage: Future generations may inherit the advantages of the economy. Social hierarchies are reinforced when children from wealthy families have better access to social capital and educational opportunities.

Challenges and Resolutions

Addressing the connected issues of social hierarchies and economic inequality is a challenging task:

Governments can implement policies such as progressive taxation, social safety nets, and level-playing-field educational initiatives to reduce income inequality.

Anti-Discrimination Measures: To attain greater equity, anti-discrimination measures based on race, gender, and other social hierarchies must be implemented.

Education and Awareness: Better understanding and social change can come from raising awareness of the impacts of economic inequality and social hierarchies.

Corporate Responsibility: Employers can assist by putting in place fair hiring practices, closing pay disparities, and promoting inclusivity and diversity.

Pursuing fairness

Economic inequality and social hierarchies are not distinct issues, but rather are entwined and mutually reinforcing. To address these problems, a multimodal approach that addresses the underlying social hierarchies and economic disparities is required. The ultimate goal is to create more equal societies where all people, regardless of social or economic background, have equal access to opportunities and resources. A commitment to social justice and a more inclusive and equitable world for all will motivate people to work together toward this goal, which will require cooperation at the individual, community, and policy levels.

8.2 Environmental Hierarchies and Sustainable Development: A Confluence of Preservation and Progress

Due to its intricate systems and finite resources, the environment is crucial to achieving sustainable development. The maintenance of the delicate balance of the natural world depends on environmental hierarchies, which include species, ecosystems, and ecological processes. Understanding the relationships between environmental hierarchies and sustainable development is essential to promoting human-environment harmony.

Environmental Structure: Levels of Ecosystems

Environmental hierarchies are the arranged levels and connections seen in ecosystems.

They are diverse in scale, ranging from individual organisms to entire ecosystems. These hierarchies are essential for comprehending the complexities of ecology and natural science.

Levels of the Environmental Hierarchy:

Individuals and Populations: The lowest level is made up of individual organisms, such as plants and animals, as well as the populations that comprise them. It examines interactions between different species as well as within the same species.

Communities: At this level, ecologists look into the composition of different species and how they interact with one another within a given habitat or ecosystem.

Ecosystems are the collective term for the biotic *(living things)* and abiotic *(non-living elements)* components of a given region, such as forests, wetlands, or oceans.

Biomes: Biomes are large ecosystems with similar flora and temperatures. A few examples are tundras, rainforests, and deserts.

Sustainable Development: Striking a Balance Between Human Needs and Environmental Preservation

Meeting present needs without jeopardizing the ability of future generations to meet their own is the aim of sustainable development. It means finding a balance between economic expansion, environmental protection, and social justice.

The concept of sustainable development recognizes that environmental hierarchies and ecosystems are the foundation of all human endeavors and that maintaining them is crucial to ensuring long-term well-being.

The connection between environmental hierarchies and sustainable development

Resource Management: Sustainable development requires responsible resource management at every level of the environmental hierarchy. This means protecting biodiversity, maintaining ecosystems, and using natural resources efficiently.

Preservation of Biodiversity: The variety of species within environmental hierarchies determines the robustness and stability of ecosystems. Sustainable development works to protect and restore biodiversity in order to preserve the health of ecosystems.

Services Offered by Ecosystems: A few of the essential tasks carried out by ecosystems are pollination, clean air, and clean water. By recognizing the value of these services, sustainable development seeks to protect them.

Climate Change Mitigation: Addressing climate change, a significant global environmental concern, is a crucial component of sustainable development. Reducing greenhouse gas emissions, converting to renewable energy, and implementing climate adaptations are all necessary for this.

Challenges and Strategies

Sustainable development and environmental hierarchies are hampered by a number of factors:

Ecosystems and sustainability may be negatively impacted by overuse of natural resources. The desire for quick financial gain is usually the driving force behind this overuse. One tactic is to employ sustainable practices in agriculture, forestry, and fisheries.

Environmental hierarchies are negatively impacted by pollution and habitat degradation. Among the tactics are stringent environmental regulations, habitat restoration, and responsible waste management.

Human Population Growth: As the world's population increases, so does the demand for natural resources. Among the tactics are family planning, education, and sustainable urban development.

Access Inequity: Ensuring equitable access to resources and the benefits of sustainable development is essential. Promoting social equity and addressing social hierarchies are the two main objectives of strategies.

An all-encompassing approach to sustainability

Environmental hierarchies and sustainable development are intertwined aspects of our shared future. Sustaining biodiversity, preserving the environment, and improving human well-being while respecting the intricate webs of ecosystems are all necessary to achieve sustainability. It necessitates an all-encompassing approach that addresses the complexities of environmental hierarchies and seeks to reconcile human needs with the imperative of preserving the natural world. Only by making such a comprehensive effort will we be able to discover the path to a more equitable, sustainable, and peaceful coexistence with the environment.

8.3 Challenging Hierarchies to Make the World Fairer: Social Justice Methods

There have always been hierarchies in human history, whether they are based on racial, gender, or economic criteria. Often, these hierarchies result in unequal power, resources, and opportunity distribution. To create a more egalitarian society, these hierarchies must be challenged and destroyed. This endeavor requires comprehensive approaches that advance social justice, reduce inequality, and promote inclusivity.

Being Aware of the Impact of Hierarchies

Hierarchies can give rise to significant social and economic disparities, with some individuals or groups enjoying greater privileges and access to resources while others are disadvantaged on a systemic level.

Access to Professional and Educational Opportunities Is Unequal: People at the top of hierarchies often have more access to these opportunities, which limits the social mobility of those at the bottom.

Reinforcement of Discrimination: Hierarchies have the potential to support discrimination based on factors such as age, gender, race, or other characteristics. Discriminatory practices impede the attainment of full equality for marginalized groups.

Techniques to Manage Hierarchies:

One of the most important first steps in addressing educational hierarchies is ensuring equitable access to high-quality education. This means fixing disparities in educational resources, boosting inclusive curricula that reflect a variety of perspectives, and boosting funding for schools.

Policies that uphold the ideas of affirmative action and equal opportunity can help marginalized groups have more equitable access to opportunities. These rules aim to eliminate historical discrimination and enhance representation.

Economic Reforms: The primary objectives of economic reforms should be equitable labor practices, minimum wage adjustments, progressive taxation, and other policies that lessen income inequality. It is possible to weaken economic hierarchies by supporting measures that promote equitable wealth distribution.

Legal Defenses: Having legal protections against harassment and discrimination based on age, gender, race, and other characteristics is essential. Strict enforcement of these laws is necessary to ensure equal rights and opportunities for all.

Initiatives for Inclusion and Diversity: Institutions and groups should work hard to achieve these objectives. Programs like diversity training, recruitment drives, and inclusive work environments must be put into place.

Cultural Awareness and Education: People can recognize and face their own prejudices when they have greater access to and understanding of cultural awareness and education. It fosters empathy and comprehension at all hierarchical levels.

Community Engagement: Local change can be achieved by addressing hierarchies through grassroots organizations and community engagement. Providing marginalized communities with the means to defend their needs and rights is a crucial strategy.

Media and Representation: It is critical to promote diverse representation in the media, entertainment, and advertising in order to combat prejudice and promote inclusivity.

Challenges and Resistance:

Opposition to Change: Those who stand to gain financially from the status quo as well as organizations usually oppose any changes to hierarchies. It will take a lot of work to overcome this resistance.

Implicit Bias: Deeply ingrained in society, implicit biases are often difficult to overcome. Addressing and increasing awareness of these biases will require ongoing effort.

Institutional Inertia: Businesses, government agencies, and labor unions may exhibit a reluctance to change. To overcome institutional inertia, tenacious advocacy and legislative action may be necessary.

Intersectionality: It's important to recognize that individuals may encounter various types of discrimination; this is referred to as intersectionality. It is advised to use intersectional methods that consider the compounding effects of hierarchies.

The Global Perspective:

Breaking down hierarchies is a global problem. It is essential to have global cooperation, shared values of justice and equity, and an awareness that hierarchies are not exclusive to any one nation or region. Global efforts to address structural, racial, and economic inequality have the potential to yield positive results.

Towards equity and justice

Addressing hierarchies will need consistent work from individuals, communities, and institutions if we are to see a more equitable world. The goal is to create a society where all people, regardless of background or identity, have equal access to opportunities, resources, and power. Despite the possible challenges along the way, creating a world where justice, inclusivity, and fair play for all are valued is an important project. As we collaborate to address hierarchies, we get closer to a just and equitable world for the generations that are here and those that will come after.

Chapter 9:

Defining New Hierarchies and Power Structures for Social Change

Alternative hierarchies offer new opportunities for societal change while challenging existing power structures. It recognizes that hierarchies are not inherently harmful while simultaneously challenging the structures that support inequality and marginalization. Alternative hierarchies can aid in our vision and aspiration of a fair and just society.

Understanding Various Hierarchies

Models of organization and decision-making called alternative hierarchies try to distribute power and responsibilities among participants more equitably. They're also called horizontal or non-hierarchical structures. These structures prioritize inclusion, shared leadership, and cooperative decision-making. They offer an alternative to the top-down, authoritarian hierarchies that are typically found in traditional institutions.

Crucial Features of Various Hierarchies:

Participatory Decision-Making: Participants have a direct say in decision-making processes, and reaching consensus is frequently valued.

Equality and Inclusivity: Alternative hierarchies aim to lessen power disparities and promote inclusivity while respecting the diverse opinions of all participants.

Collective Leadership: Leadership is usually delegated to a number of individuals or alternates in order to prevent a small group of people from possessing all the power.

Open Communication: Transparency and open communication are essential to promoting the unrestricted flow of ideas and information.

Shared Responsibilities: Participants split up the leadership duties and project- or organization-related workload.

Other Uses for Hierarchies:

Activism and Social Movements: A lot of social movements employ alternative hierarchies in order to promote inclusivity and democracy.

Occupy Wall Street and Black Lives Matter are two examples of such movements.

Worker cooperatives: Workers in worker cooperatives collectively own and manage a business, using democratic processes for making decisions.

Nonprofits and NGOs: Some nonprofit organizations use alternative hierarchies to ensure that power is distributed more fairly among their members.

Community Organizations: Alternative hierarchies are a common tool used by grassroots community organizations to empower residents and foster community development.

The Position of Technology:

Alternative hierarchies are now feasible and simpler to establish thanks to technological advancements. Online platforms, social media, and collaborative software have created new tools for decentralized decision-making and cooperative organization. Through virtual spaces, people from different geographical locations can participate in decision-making processes.

Inquiries and Comments:

Efficiency: Some argue that traditional top-down hierarchies may be more effective than alternative hierarchies, especially in large organizations.

Opposition to Change: The adoption of new hierarchies may be met with resistance from established power structures, and participants may find it difficult to adapt to the new procedures for making decisions.

Achieving True Inclusivity Can Be Difficult: Because some voices may be heard more loudly than others, there may be disparities in inclusivity.

Social Transformation and Shift:

Alternative hierarchies are crucial for social change because they upend the status quo and encourage inclusivity. They empower marginalized groups by giving previously unheard voices a forum to engage in decision-making. In this way, they combat systemic injustices and promote justice.

Concluding Remarks: Creating a More Just Society

A way to alter power relations and create a just and equitable society is through alternative hierarchies. Though they are not without challenges, they have a great deal of potential for inclusivity and social change. People and organizations that experiment with different hierarchies are part of a larger movement that seeks to overthrow oppressive systems and establish new, more egalitarian ones. We get closer to a society where power is more widely distributed and social change is guided by justice, inclusivity, and group decision-making when we accept alternative hierarchies.

9.1 Going Beyond Traditional Power Structures: Reevaluating Governance and Leadership

Traditional power structures, often characterized by hierarchical and authoritarian models of governance, are under threat in today's world. As societies become more diverse, intricate, and interconnected, there is a growing realization that traditional power structures may not be the most effective way to address today's problems. This has compelled us to reevaluate our approaches to leadership, self-governance, and self-organization, with a focus on inclusive, adaptable, and cooperative tactics.

The Difficulties of Traditional Power Structures

Marginalization and Inequality: When inequalities are exacerbated by hierarchical power structures, some people or groups may experience marginalization. Often, decisions are made with the benefit of a select few at the expense of the majority.

Rigidity: Conventional power structures often exhibit rigidity and a slow rate of condition adjustment. Bureaucracy and red tape can hinder innovation and flexibility.

Lack of Inclusivity: Because decision-making is often concentrated in the hands of a small number of individuals, diverse viewpoints and voices are frequently excluded from it. This lack of inclusivity could lead to social unrest and less-than-ideal outcomes.

Novel Approaches to Governance and Leadership:

A democracy that emphasizes public participation in decision-making is known as a participatory democracy. Through processes like open forums, citizen assemblies, and referendums, it enables citizens to take a more active role in governance.

Collaborative Leadership: Group decision-making and shared accountability are highly valued in collaborative leadership models. Leaders cooperate and co-create with their teams rather than working above them.

Networked Organizations: Networked organizations are built on the principles of flexibility and adaptability. Their organizational structures are often more flexible, encouraging cross-functional cooperation and adaptability.

Digital Platforms: With the rise in popularity of social media and digital platforms, it is now possible to organize and inspire people for social and political change in a more decentralized and grassroots manner.

Examples of Progressive Models

Participatory budgeting: This approach gives local residents direct control over the allocation of public funds in cities across the globe, ensuring that government spending is in line with community priorities.

In *"**Reinventing Organizations**,"* by Frederic Laloux, the three key concepts of teal organizations are wholeness, self-management, and evolutionary purpose. They are in favor of a more organic and decentralized process for making decisions.

Deliberative Democracy: Models of deliberative democracy encourage citizens to engage in deliberate, unbiased discussions before making decisions. They aim to raise the standard of public conversation and produce more thoughtful, inclusive policy.

Challenges and Considerations:

Complexity: It can be challenging to implement new leadership and governance models, and it may be necessary to make significant changes to institutional structures and cultural norms.

It's critical to strike a balance between autonomy and accountability, whether for oneself or a community. Ensuring that decentralized decision-making processes are transparent and considerate of the interests of all stakeholders is a challenging task.

Opposition to Change: Maintaining the status quo is often in the interests of conventional power structures. Opposition to change can seriously hinder the adoption of new models.

The Position of Technology:

In particular, the internet and digital communication tools have sparked a dramatic overhaul of governance and leadership. It has enabled greater citizen participation, real-time information sharing, and effective and quick mass mobilization.

Transitioning to More Adaptable and All-Inclusive Frameworks

The demise of traditional power structures is a response to our environment's increasing complexity and shifting needs. Although these models have limitations, they also offer chances to improve inclusivity, flexibility, and responsiveness. They acknowledge that decisions can be influenced by a variety of perspectives and that leadership and power can be distributed more widely. As societies and organizations continue to explore these alternative models of leadership and governance, they are moving closer to creating more flexible, equitable, and inclusive systems that are better equipped to address the pressing issues of our day. By rethinking leadership and governance, we can work toward a future that embodies the ideals of justice, collaboration, and personal empowerment.

9.2 Anarchism, Grassroots Movements, and Participatory Decision-Making: An Idea Convergence

Anarchism, grassroots movements, and participatory decision-making processes share values and principles that promote decentralized, bottom-up approaches to social change and governance. Though each of these concepts has unique characteristics of its own, they often share a commitment to inclusivity, nonhierarchical structures, and people taking an active role in shaping their own lives.

Grassroots Movements: People-Powered Change

Typically, ordinary people, rather than recognized authorities or institutions, initiate and lead grassroots movements, which are localized group initiatives. These movements address a wide range of social, political, and environmental issues and promote change through people-powered strategies and grassroots initiatives.

Essential Components of Community-Based Movements:

Local Empowerment: Through grassroots movements, people are given the power to directly affect decisions that affect their lives and to take action.

Community-Based: These movements are rooted in the local community and address issues that directly affect people's quality of life, such as social justice, economic equity, and environmental protection.

Decentralization: Grassroots initiatives are usually more flexible and adaptive because they are decentralized and do not usually follow hierarchical structures.

Participatory Democracy: Many grassroots movements make use of the principles of participatory democracy, which encourage people to voice their concerns, participate in decision-making processes, and collaborate to find solutions.

Participatory Decision-Making: Expanding the Voice of the Many

"Participatory decision-making" is a process that emphasizes how directly individuals or groups can impact decisions that affect them. Rather than supporting the conventional hierarchical approach to decision-making, this approach promotes diversity, collaboration, and reaching common ground.

Crucial Elements of Joint Decision-Making:

Inclusivity: Decision-making processes should be open to all parties to ensure that a range of opinions are heard and taken into consideration.

Building Consensus: Rather than relying solely on majority rule, participatory decision-making seeks to achieve consensus, where decisions are accepted by all or nearly all participants.

Transparency: Open communication of facts and discussions during the decision-making process is essential to fostering trust and accountability.

Empowerment: Participatory decision-making gives individuals the direct ability to shape policies and decisions.

Anarchism: Decentralized Governance and Opposition to Authoritarianism

Anarchism is a political theory that supports the development of decentralized, non-coercive organizational structures in place of authoritarian and hierarchical ones. Despite being misunderstood as chaotic or lawless, anarchism emphasizes self-governance and group decision-making without the need for central authorities.

Crucial Components of Anarchism:

Decentralization: Instead of centralized authority, anarchism encourages the development of small, autonomous communities.

Voluntary Cooperation: Anarchism promotes voluntary cooperation between individuals and groups in place of the use of force and coercion as a form of government.

Direct Action: Anarchists commonly employ direct action, such as civil disobedience and mutual aid, to achieve their goals and topple repressive institutions.

Non-Hierarchical Organization: Anarchist groups usually operate on non-hierarchical, consensus-based principles and place a high value on individual liberty and collective decision-making.

Meeting of Principles: Grassroots Movements and Anarchism

Because they both promote decentralized governance, nonhierarchical decision-making, and individual empowerment, anarchism and grassroots movements frequently clash. Grassroots movements represent the ideals of participatory democracy, where community members work together to address pressing issues. In favor of local empowerment, they often reject established institutions and authorities. In this sense, they are consistent with the anarchist goal of reducing centralized power.

The strategies employed by grassroots movements—such as direct action and mutual aid—reflect the anarchist principles of encouraging voluntary cooperation and opposing repressive institutions. Anarchism and grassroots movements share a vision of a society in which many people share power and decisions are made collectively with the consent of those affected.

An inclusive, decentralized governance model

Grassroots movements, participatory decision-making, and anarchism offer alternative visions of governance and social change that emphasize inclusivity, decentralized organization, and the direct participation of individuals in shaping their own destinies. Though they may not always recognize it, these movements usually share ideas with anarchism's critique of hierarchy and authority. Collectively, they contribute to a broader conversation about how society can be organized to be more just, equitable, and inclusive—where authority is shared among many and decisions are made collaboratively to address the most important issues of the day.

9.3 Models of Community Governance and Indigenous Leadership: A Fusion of Tradition and Expertise

Indigenous peoples all over the world have developed unique forms of leadership and community governance that are a reflection of their strong spiritual, cultural, and ties to the land. These models strongly emphasize group decision-making, upholding ancestral customs, and coming to agreements. While customs vary among indigenous communities, there are some universal themes that highlight the importance of long-term, comprehensive approaches to leadership and governance.

Key Tenets of Indigenous Models:

Indigenous leadership is seen in a broad sense and goes beyond particular role models. A common belief is that leadership is a shared responsibility involving participation from a range of individuals, such as elders, spiritual leaders, and members of the community in decision-making and general well-being.

Spiritual and Cultural Connection: Indigenous leadership models often include spiritual and cultural elements. Maintaining a close relationship with the land, the ancestors, and the spiritual traditions that guide the community is expected of leaders.

Consensus and Inclusivity: When making decisions, indigenous communities typically employ consensus-building techniques. Elders and community members collaborate to have conversations and make decisions that consider multiple points of view.

Interdependence with Nature: Indigenous governance is based on the notion that people are a part of nature and that safeguarding the land and its resources is crucial. Sustainability and environmental respect are essential.

Examples of Indigenous Models

Indigenous communities frequently have elder councils, which serve as a source of wisdom and guidance. These councils play a crucial role in decision-making and conflict resolution.

Clan Systems: Indigenous peoples often live in clan systems where different clans share responsibilities for leadership and governance. Each clan may have distinct responsibilities and skill sets.

Tribal Councils: In certain circumstances, indigenous nations have established tribal councils, which bring together delegates from different clans or communities to make decisions collectively.

Customary Practices: Rituals, ceremonies, and storytelling are the cornerstones of Indigenous governance. They preserve traditional values and offer guidance on decision-making.

Challenges and Resilience:

Among the many challenges faced by indigenous leadership and community governance models are marginalization, outside influences, and colonial legacies. However, many indigenous communities have shown themselves to be extraordinarily resilient in preserving and adapting their governance frameworks to meet contemporary challenges.

The Purpose of Personal Decision Making:

The idea of self-determination is crucial for indigenous peoples. It includes the right to self-govern, territorial defense, and self-determination. The United Nations Declaration on the Rights of Indigenous Peoples recognizes the importance of self-determination in preserving indigenous governance.

Sustainability and Conservation Spearheaded by Native Americans:

Ideals of sustainability and conservation are often aligned with indigenous leadership models. When it comes to initiatives to protect the environment and cultural heritage, indigenous communities have taken the lead. They possess the skills and expertise necessary to manage their lands in a way that conserves ecosystems, fosters biodiversity, and advances the welfare of coming generations.

An abundance of wisdom

A multitude of information is available from indigenous leadership and community governance models, which highlight the significance of inclusive, sustainable, and comprehensive approaches to decision-making. These models highlight the interconnectedness of people, the environment, and spirituality. Indigenous communities continue to defend their ancestors' customs and assert their right to self-governance in the face of challenges. Indigenous models provide valuable insights and lessons for creating more equitable, inclusive, and sustainable societies in a world where many people are searching for alternative forms of government that prioritize the welfare of the environment and the needs of people. Acknowledging and honoring these role models is not only a matter of justice but also an opportunity to learn from long-standing traditions.

Chapter 10:

Organizational Futures: Adapting to a Changing Environment

After being common in many areas of human society, hierarchies are radically changing in the twenty-first century in response to new opportunities and difficulties. Corporate hierarchies, governance systems, and social organizations are changing into new hierarchies as a result of globalization, rapid technological advancement, and a growing emphasis on sustainability and inclusivity.

An Introduction to the Conventional Hierarchy

In hierarchical structures, power and decision-making authority were traditionally allocated along a specific chain of command. Businesses, governments, and even social organizations made extensive use of this model. Despite providing uniformity and clearly defined hierarchies, it was often criticized for promoting inequality, quashing creativity, and disregarding the diverse needs of individuals and groups.

The Changing Landscape: Technology and Globalization

The digital era and the introduction of the internet have drastically altered traditional hierarchies. With more information available than ever before, individuals and grassroots movements are able to topple long-standing power structures. The ability to connect and collaborate globally has blurred the lines between authority because ideas and innovations can originate from anywhere.

Significant Trends Affecting the Future of Hierarchies:

Decentralization: This portends well for hierarchies in the future. Decision-making and power are shifting from the top echelons of government and institutions to a broader range of stakeholders, such as employees, citizens, and international networks of people.

Networked Organizations: Linked individuals and groups collaborating toward a common goal make up networked organizations, which pose a threat to the traditional hierarchical model. These businesses value adaptability, inclusivity, and agility highly.

Diversity and Inclusivity: Due to the recognition of various viewpoints, hierarchies have grown to be more inclusive. Gender, racial, and cultural diversity are becoming recognized as assets, and institutions of higher learning are making efforts to become more welcoming.

Sustainability and Responsibility: Given the urgent need to address global issues like social inequality and climate change, hierarchies need to be reevaluated. Policies are being implemented by governments to address social and environmental issues, and businesses are beginning to place a greater emphasis on sustainability and corporate social responsibility.

Digital and Technological Innovation: As technology advances and continues to reshape hierarchies, new forms of communication, decision-making, and economic models become possible. For example, the accountability and transparency of hierarchies could be completely transformed by blockchain technology.

Examples of New Hierarchy Models

Organizational structures that are flatter and encourage open communication, teamwork, and worker empowerment are replacing strict hierarchical structures in many businesses.

Participatory Governance: Some governments are looking into models of participatory governance that involve the public in the decision-making process through town hall meetings or online forums.

Cooperative Models: Worker cooperatives, in which employees jointly own and run a business, are becoming more and more popular as an alternative to traditional hierarchical structures that place a premium on shared ownership and decision-making.

Challenges and Disagreements:

Hierarchies face several challenges and conflicts as they shift:

Striking the Right Balance Between Inclusivity and Efficiency: While excessive decentralization can lead to inefficiency, excessive hierarchy can stifle creativity and engagement. It can be challenging to strike the ideal balance between these two elements.

Adapting Cultural Norms: To modify cultural norms and expectations about hierarchies, organizational and societal mentalities may need to shift. Opposition to this process might exist.

Transparency and Accountability: It can be challenging to preserve transparency and accountability in decentralized structures as decision-making processes grow more dispersed.

Last Words: An Incendiary Journey

Subsequent hierarchies will be drastically transformed in the name of sustainability, adaptability, and inclusivity. Traditional power structures are being redesigned in order to better address the complex issues of our day, such as environmental crises and social inequality. As they navigate this shifting landscape, governments, corporations, and individuals can reap the rewards and overcome the challenges of new hierarchy models. The future form of hierarchies will depend on the creation of more dynamic, inclusive, and accountable systems that are better suited to the demands of the twenty-first century. By doing this, we may discover innovative approaches that optimize human potential and promote the welfare of both people and the environment.

10.1 Innovation and Resilience in the Twenty-First Century: Adjusting to a Changing World

In the twenty-first century, nearly every aspect of human existence has experienced unprecedented change and transformation. In addition to constant technological advancement, people and societies also need to adapt to shifting social and environmental conditions. Adapting to a changing environment is not only essential, but it's also a critical mindset and skill that enables us to welcome change, deal with uncertainty, and thrive in the face of uncertainty.

Change Is Occurring More Quickly:

In the annals of human history, no era has undergone as much rapid change as the 21st century. Because of rapid technological advancement, globalization, and changing social mores, the status quo is never stable. The COVID-19 pandemic, for instance, illustrated the necessity of responding quickly and nimbly to unforeseen challenges.

Principal Drivers of Change:

Technology: Developments in artificial intelligence, biotechnology, and renewable energy are changing jobs and our way of life.

Globalization: The interdependence of economies, cultures, and societies has altered how we interact with one another and conduct business.

Environmental Concerns: The need for sustainable practices and the effects of climate change are causing industries and policies to change globally.

Demographic Shifts: Aging populations, urbanization, and changing family structures are affecting social norms and economic dynamics.

The Benefits of Adjustability

Adaptability is a basic skill in a world that is changing. It means responding to changes as well as actively seeking out opportunities for growth and innovation. One's ability to adapt is essential for success in both the personal and professional domains as well as for societal resilience.

Individual Modification:

Continuous Learning: Pursuing lifelong learning is essential. Over the course of their lives, people must be open to learning new things.

The ability to overcome challenges and cope with uncertainty is a valuable personal attribute that is often associated with resilience and emotional intelligence.

Flexibility: The capacity to change course when called for characterizes those who are adaptable.

Creativity and Innovation: The capacity to think creatively and develop original solutions is becoming increasingly valuable in the dynamic job market.

Cultural Adaptation:

Policies that Promote Inclusivity: Governments must put in place policies that promote inclusivity in order to mitigate the effects of changes on marginalized communities.

Sustainable Practices: To solve environmental issues, industries and communities must implement sustainable practices.

Technology and Education: Ensuring equitable access to technology and education for all individuals is essential to achieving fair adaptation.

Healthcare Resilience: When a medical emergency arises, the system must be prepared to respond and adjust as needed.

The Challenges of Adjustment:

Opposition to Change: Common causes of resistance to change include fear, uncertainty, and attachment to the status quo. Convincing people and organizations to adapt can be challenging.

Overwhelm: The quick speed of change can lead to burnout and stress. Well-being and adaptation need to be balanced, which is an important concern.

Inequities: People differ in their ability to thrive in a changing environment because they have different access to the opportunities and resources needed for adaptation.

Embracing Shift:

Creating a Growth Mindset: By highlighting the notion that aptitude and intelligence can be developed, a growth mindset encourages a positive outlook on learning and adaptation.

Developing Resilience: The capacity to bounce back from adversity and overcome challenges can be effectively fostered through self-care, social support, and mindfulness practices.

Collaboration and Innovation: Fostering creative environments and working together with others can speed up adaptation, whether in a professional or personal context.

Education and Awareness: By making people more aware of the importance of adapting as well as by giving them access to resources and education, communities and individuals can be given the tools they need to prosper in a changing global environment.

Closing Reflections: confidently sailing into the future

It's not only about getting by in a constantly changing environment; it's also about seizing opportunities for development and progress. In the twenty-first century, adaptability and success rely on accepting change on a personal, societal, and global scale. As the world changes, those who can manage uncertainty with innovation, adaptability, and an open mind will be well-positioned to thrive in the face of change and help create a more inventive and resilient future. Rather than just being a response to a shifting environment, adaptation is a proactive strategy for creating a better tomorrow.

10.2 The Vital Conflict: Balancing Hierarchy and Coordination

Finding a balance between hierarchy and cooperation is one of the main problems with the way human societies, institutions, and relationships are structured. While hierarchy provides structure, order, and accountability, cooperation fosters inclusivity, creativity, and flexibility. Achieving the optimal equilibrium between these two elements is crucial for promoting societal unity, effective governance, and personal well-being.

The Organizational Hierarchy:

People or things are arranged in hierarchical systems of organization according to their authority or significance. It is usually identifiable by a clear chain of command, roles and responsibilities, and the authority that those at the top have to make decisions. Hierarchies come in a variety of forms, including those seen in governmental structures, business organizations, and social and familial relationships.

Benefits of a Hierarchy:

Clarity and Order: A hierarchical structure provides a clear structure that helps people understand their roles, responsibilities, and accountabilities better. This clarity may lead to efficient task allocation and decision-making.

Accountability: Hierarchies promote accountability because individuals are held responsible for their actions within the parameters of their designated roles.

Efficiency: Hierarchy can occasionally aid in more effective task coordination and execution. This is particularly true for organizations where it is essential to have clearly defined processes and procedures.

The Negative Aspects of Structures

Rigidity: Too much hierarchy can lead to rigidity, which hinders innovation and makes it challenging to adapt to changing circumstances.

Inequality: Hierarchies can support power disparities, in which individuals at the top have an excessive amount of influence and control.

Silos and Communication Barriers: In highly hierarchical organizations, communication barriers frequently lead to silos and a lack of information sharing.

The Nature of Cooperation:

Cooperation is the act of people or groups working together to solve issues or achieve common goals. It is characterized by collaboration, group decision-making, and an openness to hearing and weighing various viewpoints. Both inside and outside of hierarchical structures, collaboration is possible.

The Advantages of Teamwork:

Collaboration is an ideal approach to solving complex, multifaceted problems because it promotes inclusivity and the integration of diverse points of view.

Innovation: Collaborative environments often foster creativity and innovation because individuals bring a variety of skills, experiences, and backgrounds to the table.

Adaptability: Collaboration facilitates the capacity to change course and overcome unanticipated challenges.

The Negative Effects of Teamwork:

Lack of Accountability: When cooperation is overemphasized, it can be hard to hold people accountable for their actions or decisions and accountability can become diffused.

Decision paralysis and conflict: Since decisions in cooperative settings have to be made by consensus, there is a chance that disagreements will arise and that making decisions will take longer.

Inefficiency: Cooperation may not always be the best course of action when there are too many parties involved in the process and prompt decisions are required.

Balancing Hierarchy and Collaboration:

The key to effectively managing societies, groups, and interpersonal relationships is striking the right balance between cooperation and hierarchy.

Here are a few strategies for achieving this balance:

Contextual Approach: Recognize that there may be several approaches to achieving the ideal harmony between hierarchy and cooperation, depending on the circumstances. While long-term planning may benefit more from collaborative processes, a crisis situation may necessitate more top-down decision-making.

Effective Communication: Within hierarchical structures, promote open and effective communication to avoid silos and ensure that information flows freely.

Transparent Decision-Making: Encourage transparency in decision-making processes within hierarchies in order to maintain accountability and cultivate trust.

Collaborative Leadership: By listening to a variety of perspectives and involving others in decision-making, leaders can apply a collaborative leadership strategy in hierarchical environments.

Adaptive Structures: By creating flexible, adaptive structures inside hierarchical frameworks, organizations are able to respond to changing circumstances while maintaining accountability.

The Art of Equilibrium

Finding a balance between hierarchy and cooperation is a challenging and ongoing task in the governance of societies, organizations, and relationships. Every component has benefits and drawbacks, and the ideal balance may vary based on the specific circumstances. Attaining this balance requires careful consideration, adaptability, and a willingness to change as circumstances do. To effectively manage this tension and build environments that are inclusive and structured, efficient and creative, and accountable and flexible, requires a unique set of skills. This balance is the key to successful and peaceful cooperation and governance in the modern world.

10.3 Towards Sustainable and Inclusive Hierarchies: A Framework for 21st-Century Governance

Building inclusive and sustainable systems of governance requires rethinking hierarchies in an era marked by global challenges like social inequality, environmental degradation, and climate change. Traditional hierarchies have often faced criticism for increasing inequality and disregarding environmental concerns. Hierarchies can be changed and adjusted while retaining their organizational benefits to address these pressing issues.

The Evolution of Hierarchy Interpretation

Traditionally, hierarchies have been associated with centralized power structures, often characterized by top-down decision-making and limited inclusivity. Conversely, the concept of sustainable and inclusive hierarchies advocates for a more adaptable and equal form of governance.

Crucial Pointers for Sustainable and Inclusive Hierarchies:

Decentralization of Power: In decentralized hierarchies, assigning responsibility and decision-making authority to different levels and stakeholders is given top importance. This promotes accountability and reduces the amount of power that a small number of people hold.

Environmental Stewardship: Because they acknowledge the urgent need to address environmental issues, sustainable hierarchies place a strong emphasis on environmental stewardship and responsible resource management. They incorporate sustainability into decision-making processes and policies.

Diversity and Inclusivity: Sustainable hierarchies actively seek to advance diversity and inclusivity in decision-making processes because they understand the value of different points of view. Ensuring equal participation opportunities and elevating the voices of underrepresented groups may be necessary to achieve this.

Accountability and Openness: Transparency is the cornerstone of inclusive hierarchies. Open communication and information accessibility ensure that decisions are made with accountability in mind, which builds trust within the hierarchy.

Examples of Sustainable and Inclusive Hierarchies:

Participatory budgeting: This approach, which has been adopted by several municipalities, allows residents to directly influence the distribution of public funds. This makes it easier to ensure that choices are made in a way that promotes inclusivity and takes community priorities into account.

Teal Organizations: In ***"Reinventing Organizations,"*** Frederic Laloux claims that in the business world, teal organizations promote self-management, wholeness, and evolutionary purpose. Establishing more inclusive and sustainable workplace hierarchies is their aim.

Multi-Stakeholder Partnerships: These are organizations and initiatives that collaborate with a variety of stakeholders, including corporations, governments, and civil society, to advance sustainability and inclusivity while tackling challenging global issues.

Challenges and Considerations:

Striking the Right Balance Between Inclusivity and Efficiency: Striking the right balance between inclusivity and efficiency can be challenging. Decentralization and inclusivity may slow down the decision-making process, which could be problematic in some situations.

Cultural Shifts: In order to convert traditional hierarchies into inclusive, sustainable systems, significant cultural changes are frequently required, as well as a willingness to challenge accepted conventions and practices.

Resource Allocation: Efficient resource allocation and decision-making procedures must be developed to ensure that inclusivity does not lead to inefficiency or indecision.

The Position of Leadership:

The leaders in hierarchies have a major role in driving the change toward inclusivity and sustainability. They ought to uphold the principles of accountability, openness, and environmental stewardship; they ought to be agents of change and responsible decision-makers by example.

The Universal Principle:

Changes in governance and hierarchies are imperative in order to address global issues such as social inequality and climate change. In a globalized society where environmental and human welfare are closely related, having sustainable and inclusive hierarchies is not just a matter of taste. To govern in the twenty-first century, we need to adopt inclusive, adaptable, and sustainable models instead of the rigid hierarchies of the past.

Rethinking Governance to Create a Better Tomorrow:

The concept of inclusive and sustainable hierarchies represents a path toward a more equitable, responsible, and sustainable society. It is an acknowledgement that, as they adjust to the demands of the twenty-first century, hierarchical structures can effectively handle complex global challenges. By embracing these concepts and actively pursuing sustainability and inclusivity within hierarchies, we can move closer to creating governance structures that better satisfy the diverse needs of our societies and safeguard the environment.

Conclusion:

Conclusions and Prospects: Choosing the Next Step

While navigating life, societies, and organizations, we pick up valuable lessons that shape our perspectives and guide our actions. These lessons serve as a roadmap for the future as well as a window into the past. We examine how experience influences our objectives and choices in this examination of the lessons we've learned and our future plans, which also provides a roadmap for our future travels.

The Capacity to Adjust:

One of the most significant lessons we have learned is the value and practicality of adaptation. Since change is the only thing that is constant in life, resilience and personal growth depend on one's ability to adapt to new circumstances. For instance, the COVID-19 pandemic highlighted the importance of adaptability by forcing individuals, communities, and nations to quickly adjust to previously unheard-of challenges.

Inclusion and Diversity:

There's another crucial lesson about the value of inclusivity and diversity. Embracing a range of perspectives, backgrounds, and voices fosters creativity and broadens our collective understanding. Because of our diversity, we are able to create more just and compassionate communities and find creative solutions to complex problems.

The Value of Sustainability

It is becoming increasingly evident that environmental stewardship and sustainability are critically needed. We can learn valuable lessons about the necessity of sustainable practices and responsible resource management from the degradation of ecosystems, climate change, and resource depletion.

Modern technology

Our experiences have shown how innovation and technology can change a situation. Human creativity has proven capable of addressing global challenges, as demonstrated by the advancements in artificial intelligence and renewable energy. These tools also function as a useful reminder of our responsibility to use technology sensibly and for the good of society at large.

Globalization and Connectivity:

Unquestionably, our world is interconnected, and these lessons emphasize the necessity of finding global solutions to address global issues. In the face of cross-national economic disparities, pandemics, and climate change, global solidarity and cooperation are essential.

Transparency and Accountability:

Transparency and accountability are essential components of responsible governance. In the wake of scandals and crises, it has become increasingly important to hold individuals, organizations, and governments accountable for their actions and decisions.

The Position of Leadership:

Without strong leadership, progress and positive change cannot be accomplished. Others are greatly impacted by visionary and compassionate leadership. Take inspiration, unity, and empowerment from other leaders.

Lessons in Resilience and Adaptability:

Throughout history, people and societies have shown incredible resilience in the face of adversity. These lessons serve as a reminder of the human capacity for resilience in the face of overwhelming adversity.

Potential Paths:

The lessons we have learned from the past help to illuminate the path ahead.

Looking ahead, a number of significant trajectories and priorities become clear:

Sustainable Practices: Achieving sustainability is one of the main objectives. We must use sustainable energy sources, consume responsibly, and preserve the environment if we want to preserve our planet.

Inclusive Societies: It is crucial to encourage diversity and inclusivity in all areas of life. This means addressing disparities, getting rid of institutionalized discrimination, and giving voice to those who are marginalized.

Technological Innovation: The morally and responsibly developed and applied technology will continue to influence our future. Adopting innovation while ensuring that it benefits all of humanity is a vital course of action.

Global Cooperation: The lessons learned emphasize the necessity of global cooperation in order to address pressing challenges. When it comes to issues like public health and climate change, international cooperation is crucial.

Ethical Leadership: Ethical leadership is still very much needed. It is these leaders who prioritize honesty, integrity, and the well-being of their constituents that will continue to shape future developments.

Resilience and Adaptability: It is critical that we cultivate resilience and adaptability because we will undoubtedly face unforeseen challenges. Building people's and communities' resilience to face adversity is a crucial first step.

Accountability and Transparency: Encouraging accountability and transparency ensures that individuals are held responsible for their actions in governance, institutions, and society at large.

Constructing a more auspicious future

We have learned invaluable lessons from our shared experiences, and they serve as a compass to help us move toward a better tomorrow. The wisdom of adaptability, diversity, sustainability, and moral leadership serves as our compass for the future. If we apply these lessons to our daily work and strive for continuous improvement, we can address global issues, create more equitable societies, and ensure a sustainable future for future generations. As we move forward, we have the opportunity to symbolize this change and create a world that embodies both the future we hope to create and the lessons we've learned. Mahatma Gandhi once said, "You have to be the change you wish to see in the world."

- Overview of Key Concepts: Understanding the Complexities of Hierarchies

Hierarchies have played a crucial role in the formation of institutions and societies as systems of governance and organization throughout human history. However, as the twenty-first century gets more complex, our understanding of hierarchies is evolving. In this summary of key ideas, we'll go over the primary topics discussed in the discussion of hierarchies, including their historical development, contemporary issues, and potential for modification.

Clearly defined hierarchy.

Organizational structures called hierarchies are characterized by the arrangement of elements—people, things, or components—in accordance with their level of importance, power, or influence. Traditional hierarchies usually feature a centralized center of power and a clear chain of command.

The Role of Organizations:

Hierarchies provide structure and order to enable efficient decision-making, accountability, and responsibility distribution. They have had a major impact on governments and organizations, among other aspects of human society.

Conventional Hierarchies' Drawbacks

Despite offering stability, hierarchies have disadvantages when applied. An excess of hierarchy can lead to rigidity, inequality, and inefficiency. Communication barriers, power disparities, and resistance to change are common traps.

The concepts of evolution and adaptation

The idea of the flexibility and evolution of hierarchies is essential. Organizational structures are evolving in response to 21st-century demands. This adaptability is necessary for resilience and growth.

Sustainability and Inclusivity:

Sustainability and inclusivity are two key paradigm shifts. More and more, hierarchies are being redesigned to promote sustainability, environmental stewardship, and responsible resource management. The significance of diversity, inclusivity, and the inclusion of underrepresented voices is becoming increasingly apparent.

Connectivity and Globalization:

Our world is defined by globalization and interconnectedness. International cooperation and solidarity are required to address global issues such as pandemics and climate change. Conventional lines are getting increasingly blurry.

Leadership's Impact:

Leadership plays a crucial role in steering hierarchies toward progress and positive change. Empathic and visionary leadership has the power to empower, inspire, and unite individuals and organizations.

Both adaptability and durability:

Historical lessons have taught us about the adaptability and resilience of humans. We have seen individuals and communities overcome adversity and thrive in challenging circumstances.

Moral Lessons to Bear in Mind:

The necessity of moral leadership and the ethical and responsible use of technology have gained traction in the discourse surrounding hierarchies. Ethical considerations encompass responsibility, transparency, and the moral application of technology.

Future Points of Reference:

The lessons we have learned thus far will direct us into the future. We're moving in the direction of operations that are more ecologically friendly, diverse, creative, international cooperation, moral leadership, flexibility, and accountability.

A Vision for a Sustainable and Inclusive Hierarchy:

The ultimate objective is to create inclusive and long-lasting hierarchies. In these organizational structures, decentralized power, environmental stewardship, inclusivity, diversity, transparency, and accountability are prioritized.

The Direction to Go:

We will need to become more resilient, adopt an ethical leadership style, welcome technological innovation, encourage inclusivity, collaborate globally, and learn to adapt to change in the future. If we put these lessons into practice, we can create a future that is more equitable, sustainable, and prosperous.

In order to meet the demands of the contemporary world, hierarchies are evolving to strike a balance between structure and flexibility, innovation and tradition, efficiency and inclusivity. As we negotiate the complexities of our globalized world and confront challenges, hierarchies have the potential to positively impact things; all we have to do is adjust, accept morality, and prioritize the well-being of people and the environment. We now have the opportunity to create governance and organizational frameworks that accurately reflect the knowledge we have acquired and the kind of future we aspire to build.

- An Argument in Favor of Sustainable Hierarchies: Building an Ethical Community

It is imperative to recognize the role sustainable hierarchies will play in building a more resilient and just society in the face of unprecedented challenges. Conventional governmental, social, and organizational structures are evolving, and this change presents a fantastic opportunity for positive change. This call to action addresses the urgent need for sustainable hierarchies in addition to describing the steps required to advance a future where sustainability, inclusivity, and responsible stewardship are prioritized.

The Necessity of Sturdy Organizations

With the problems our world faces, from environmental degradation and climate change to social inequality and global crises, a new approach to governance and organization is required. Traditional hierarchies have often faced criticism for perpetuating inequality and neglecting their environmental responsibilities. However, hierarchies are adaptable and have the potential to be positive forces. In a world where collaboration and interdependence are crucial, sustainable hierarchies are not an option—they are a necessity.

Principles of a Sustainable Hierarchy:

Decentralization of Power: The existing concentration of power among a small number of people needs to be replaced with a more distributed authority in order to foster inclusivity, accountability, and flexibility. There should be more participants in the decision-making process.

Environmental Stewardship: It's critical to consider sustainability when making decisions. Reducing climate change, protecting the environment, and managing resources responsibly need to be at the top of the hierarchy.

Diversity and Inclusivity: Because diverse perspectives are valuable, hierarchies ought to make a concerted effort to advance inclusivity and diversity. Underrepresented voices should be heard, and equal opportunities for participation should be offered.

Transparency and Accountability: Open communication and clear decision-making processes are necessary to build trust and ensure that hierarchies answer to the people they serve.

An Appeal with Several Aspects:

In order to practice responsible leadership, people in positions of power must support diversity, sustainability, and moral guidance. They should act as role models for the greater community by letting these ideals guide their decisions and deeds.

Environmental Responsibilities: Whether in business, government, or civil society, environmental responsibilities must be placed at the top of all hierarchies. This means supporting conservation efforts, promoting responsible consumption, and making the switch to sustainable energy sources.

Organizations should actively engage in inclusive decision-making processes that accept feedback from a variety of perspectives. Being inclusive promotes innovative thinking and ensures that a greater variety of needs and priorities are taken into consideration.

Because many of the challenges we face are global in nature, international cooperation is essential. Cooperation on issues such as public health, economic inequality, and climate change ought to be a primary priority for hierarchies within and between nations.

Use of Technology Ethics: Technology development and application should be guided by ethical principles. Ensuring that technological advancements are applied for the benefit of both Earth and humanity is imperative for hierarchies.

Adaptive Resilience: Hierarchies should encourage resilience by emphasizing the need of adaptability and being prepared for unforeseen challenges. Giving people and communities the tools they need to endure adversity is essential.

Accountability and Transparency: Hierarchies should encourage accountability and transparency in all areas of governance to guarantee that individuals and organizations are held responsible for their actions and choices.

The Long-Term Structures' Possible Advantages:

Future prospects are bright with sustainable hierarchies. Instead of doing away with structure and order, they seek to modify these structures to meet the demands of the twenty-first century. By implementing sustainability, inclusivity, and ethical governance, hierarchies have the ability to address global concerns, foster innovation, and create more equitable and compassionate societies.

A world deserving of our endeavors

An argument in favor of sustainable hierarchies is an argument for building a society that meets the demands of the twenty-first century. It is an organizational and governing vision that prioritizes the well-being of people and the environment while acknowledging the interconnectedness of our global community. As we navigate the complexity of the modern world, we have the opportunity to create hierarchical structures that symbolize the lessons we've learned and the kind of future we hope to achieve. This call is an open invitation to collaborate in reshaping hierarchies for the benefit of a world that merits striving for—a world where sustainability, inclusivity, and ethical responsibility are not just ideals but also commonplace realities.

Bibliography

List of references and further reading: Expanding Your Hierarchy Knowledge

Understanding hierarchies requires a multidisciplinary approach that draws on ideas from psychology, ecology, sociology, history, and other disciplines.

For a deeper dive into the world of hierarchies and related topics, the following resources and additional reading materials offer insightful analyses, diverse viewpoints, and the opportunity to investigate this complex topic from multiple perspectives:

Books:

"Sapiens: A Brief History of Humankind"* by *Yuval Noah Harari: The development of human societies and hierarchies, as well as the impact of culture on the planet, are all thoroughly covered historically in this book.

Frederic Laloux's book ***"Reinventing Organizations"*** delves into the idea of self-managing hierarchies and the evolution of organizational structures in the twenty-first century.

"The Social Animal" by Elliot Aronson delves into the psychology of behavior and how social hierarchies impact our decisions and relationships.

The book **"The Origins of Order: Self-Organization and Selection in Evolution"** by Stuart A. Kauffman delves into the formation of hierarchical structures in a range of complex systems, such as biological evolution and human societies.

"The Nature of Order: An Essay on the Art of Building and the Nature of the Universe" by **Christopher Alexander:** Examine the concept of hierarchical patterns in architecture and design, and think about how it affects the welfare of people.

Academic Journals

"Nature": A reputable scientific journal covering a wide range of topics, including ecology, evolution, and the natural world.

The *"American Journal of Sociology"* offers sociological analysis of human hierarchies, power relationships, and social structures.

An academic journal called *"Ecology Letters"* focuses on ecological hierarchies, food webs, and ecosystem dynamics in order to provide insights into the natural world.

"Psychological Review": Review research on social interactions, behavior in humans, and the psychology of hierarchies.

Websites and Electronic Resources:

The Stanford Encyclopedia of Philosophy's *"Social Institutions"* resource explores the philosophical implications of social structures and hierarchies.

Visit National Geographic's *"Ecosystems"* page to learn more about ecological hierarchies, food chains, and species connectivity in ecosystems.

World Economic Forum - "The Fourth Industrial Revolution": Acknowledge how technology is transforming future work environments and social structures in society.

Visit TED Talks to see a wide range of presentations on topics like leadership, innovation, social change, and hierarchies.

Images and Documentaries:

A documentary titled *"The True Cost"* examines the fashion industry's supply chain and the ways in which labor and the environment are impacted by economic inequality.

"The Corporation" delves into the role of corporations in modern society and their hierarchical structures.

In particular, the documentary *"The Social Dilemma"* looks at how algorithmic and data-driven decision-making have been affected by social media and technology in modern hierarchies.

Radio programs:

"The Tim Ferriss Show": Well-known guests routinely discuss leadership and hierarchies as they discuss a variety of subjects pertaining to life, business, and success.

"Hidden Brain": This podcast explores how human behavior is shaped by unconscious tendencies, including how hierarchies influence our decisions.

The journey is about learning about and investigating the world of hierarchies. By interacting with these references and additional reading materials, you can acquire a deeper understanding of hierarchies in all their complexity, from the natural world to human societies, from the past to the evolving challenges of the future. Each resource offers a unique perspective, enabling readers to gain a deeper, more comprehensive understanding of this intricate and dynamic subject.

Index

- Subject and Term Index: Working Around Knowledge Hierarchies

The study of hierarchies covers a vast and intricate area, touching on many different facets of human knowledge and experience. This extensive list of topics and keywords is meant to assist you in navigating the complicated world of hierarchies. Whatever your area of interest, whether it be social structures, ecological hierarchies, historical perspectives, or contemporary trends, this index provides a guide to the key concepts and themes surrounding hierarchies.

The Ecology Hierarchies:

Trophic Levels: The hierarchy of energy transfer in ecosystems, which moves from producers to consumers, represents the flow of energy through different organisms.

Keystone Species: Organisms that are disproportionately important for maintaining the structure and functions of an ecosystem.

Biodiversity is the range of organisms that make up an ecological hierarchy, including species, genetic, and ecological diversity.

Food webs are networks that depict the complex relationships and exchanges between various species within an ecosystem.

The capacity of an ecosystem to carry on operating and maintaining its structure over time in the face of outside disturbances is known as ecosystem stability.

Class Organizations:

Domestic Hierarchies: Social structures that exist in families include gender dynamics, generational hierarchies, and parental roles.

Class Hierarchies: When a society is divided into social and economic classes, inequality is often the outcome.

Power dynamics refers to the distribution of power and influence among members of social hierarchies, including both political and economic control.

"Inequality aversion" is a psychological process that encourages people to oppose and confront social hierarchies that are marked by inequalities.

Cooperation and Competition: How social hierarchies are shaped by the interaction of cooperative and competitive factors.

Historical Points of View:

Early Human Societies: These are the hierarchical, typically egalitarian, prehistoric and early human communities.

Feudalism and Monarchies: The medieval European hierarchical systems were typified by the supremacy of feudal lords and monarchs.

States and Empires: The emergence of nation-states and empires, distinguished by complex governmental frameworks and swift geographic growth.

The Industrial Revolution refers to the social upheavals that occurred in the 18th and 19th centuries as a result of advancements in technology and shifting economic class structures.

Novel Advancements:

Through the use of technology and the internet, the digital age has redefined conventional power structures and information hierarchies.

In virtual organizations, leadership is distributed and decentralized, typical of remote work and collaboration.

Sustainability: Ethical hierarchy, conscientious resource management, and environmental sustainability are becoming more and more important in the face of global challenges.

Social Media and Virtual Communities: The impact of digital platforms on modern hierarchies, particularly in relation to virtual influence and information exchange.

Alternative Hierarchies: Analyzing non-traditional power structures such as democratic decision-making, native leadership styles, and grassroots endeavors.

Psychological Aspects:

Status and Power is the psychology of power dynamics and status-seeking in both individual and group hierarchies.

Altruism and Cooperation: In social hierarchies, the behavioral and cognitive strategies that promote altruism and cooperation.

It is the psychological components of *"inequality aversion"* that drive individuals to resist and challenge social hierarchies that are marked by inequality.

The Human Brain and Social Hierarchies aims to explore the neural bases of human hierarchical cognition and behavior.

Key Points to Consider on Ethics:

In hierarchical structures, ethical leadership and governance are based on the principles of integrity, transparency, and the well-being of constituents.

Justice and Sustainability: It is morally necessary to uphold environmental sustainability and social justice within societal and organizational hierarchies.

Accountability and Responsibility: The moral duty placed on people and institutions to accept ethical accountability for the decisions and deeds they commit to while operating in hierarchical systems.

Being an environmental steward means having a moral responsibility to protect natural ecosystems and manage resources well.

The Future of Hierarchy:

Adapting to a Changing World: In order to satisfy the demands of a dynamic global environment and a changing society, organizational structures must undergo constant change and adaptation.

Sustaining Harmony between Collaborative, Inclusive Procedures and Hierarchical Structures: This is an ongoing challenge.

Examining unconventional power structures as a way to effect positive social change is known as *"Alternative Power Structures and Social Change."*

The Future of Hierarchies: Discussion and conjecture about potential directions and modifications that hierarchies may take in the ensuing years and decades.

This index is a great resource for people who want to learn more about hierarchies and the various dimensions that they have. These topics and keywords offer a starting point for additional research and critical analysis, regardless of your interests in understanding ecological systems, analyzing historical hierarchies, understanding social structures, or forecasting future organization and governance patterns. This index is your passport to unlocking the many insights and secrets hidden within the intricate, varied, and ever-evolving realm of hierarchies.